Tower to Tower

Tower to Tower

Gigantism in Architecture and Digital Culture

Henriette Steiner and Kristin Veel

The MIT Press
Cambridge, Massachusetts
London, England

© 2020 Massachusetts Institute of Technology

All rights reserved. No part of this book may be reproduced in any form by any electronic or mechanical means (including photocopying, recording, or information storage and retrieval) without permission in writing from the publisher.

This book was set in Stone Serif and Stone Sans by Jen Jackowitz. Printed and bound in the United States of America.

Library of Congress Cataloging-in-Publication Data

Names: Steiner, Henriette, 1980- author. | Veel, Kristin, author.
Title: Tower to tower : gigantism in architecture and digital culture / Henriette Steiner and Kristin Veel.
Description: Cambridge, Massachusetts : The MIT Press, [2020] | Includes bibliographical references and index.
Identifiers: LCCN 2019031894 | ISBN 9780262043922 (hardcover)
Subjects: LCSH: Architecture--Composition, proportion, etc. | Digital communications--Social aspects. | Largeness (Philosophy)
Classification: LCC NA2760 .S76 2020 | DDC 720.1--dc23
LC record available at https://lccn.loc.gov/2019031894

10 9 8 7 6 5 4 3 2 1

For Sofus

Contents

Acknowledgments xi

Introduction 1
    Prologue: Scaffolding 1
    A Tale of Two Cities, A Tale of Two (and More) Towers 4
    Gigantism 8
    Vertical and Horizontal Gigantism 13
    Gigantic Dichotomies, Containers, and Leaks 18
    Tower to Tower 20
    Latent Gigantism 23

1   The Eiffel Tower: Grand-Scale Montage and Lightning Rod for Meaning 28
    Prologue: An Elevator Ride 29
    Going Up 32
    Montage as Method 35
    Seeing the Eiffel Tower See: Expanding Meaning 47
    Leaks, Crisis, and Gigantism 53

2   The Twin Towers: The Remanence of the Twins 64
    Prologue: A Telephone Conversation on 9/11 65
    Scales of Impact 68
    The Twin Towers and the Intermingling of Vertical and Horizontal 74
    Challenging the View from Above 80
    Is Architecture in or above History? 86
    Towering, Falling 94

3   The One World Observatory: Caught between Vertical and Horizontal Gigantism 100
    Prologue: Gigantic Buildings 101
    The Sky's the Limit 104
    One World Trade Center and the New Metropolitan Mainstream 109

The Womb as Time Machine   116
A Leaking Container   124
New Visual Orders   129
Temporal and Material Excess   136

4 The Ground Zero Site: Calmly Common   142
Prologue: Trump Tower   143
The Fifth Leak   146
Never Ever Modern   149
Calculated Publics at the Edge of the Memorial Pools   152
From Containment to Leaks   164

5 Into the Ground: Cataloging Latent Gigantism   172
Prologue: The Earth Leaks   173
Tower to Tower and Beyond   175
Kool Chinese Twisted Twin Towers   181
Telling Tales of Tech Giants   188
Big Business beneath Manhattan   195
Epilogue: The Bias Cut   202

Notes   205
Index   223

Maria Finn, *Unfinished #19*, pencil on paper, 29 × 42 cm, 2018. © Maria Finn.

Acknowledgments

This book is the latest outcome of our long-term collaboration around a number of articles, edited books, conferences, seminars, travels, and many more challenging, weird, fun, rich, tiring (especially for the people around us), and unexpected discussions, disruptions, and digressions than we ever thought would be necessary to precede the writing of this book—conversations that go back more years than we feel ready to recall. We would like to thank the Independent Research Fund Denmark; the Ministry of Higher Education and Science, Denmark; the Swiss National Science Foundation; the Department of Urban Studies and Planning at Massachusetts Institute of Technology; and Churchill College, Cambridge, as well as our respective departments at the University of Copenhagen and the whole community around the Uncertain Archives research group for encouraging our work at key moments in the process. The effort of a much greater number of institutions and people than can be mentioned here has gone into the production of this book, and we are grateful for the support we have received throughout. Moreover, we would like to thank a number of friends and colleagues including Daniela Agostinho, Lene Asp, Peter Carl, Natalie Gulsrud, Alison McDougall-Weil, David Midgley, Maximilian Sternberg, Nanna Bonde Thylstrup, Frederik Tygstrup, Andrew Webber, as well as the three anonymous reviewers for their readings and thoughtful feedback on the developing manuscript ideas. We also would like to thank Pamela Siska from MIT's Writing and Communication Center, Rachel Malkin and Merl Storr for their courageous help with language editing of wild chapter drafts, as well as Kristen Van Haeren for her help with image selection. At the MIT Press, we are grateful to our editor Doug Sery for all his support and encouragement and would like to thank Noah J. Springer and the rest of the production team for their thorough and thoughtful work on our manuscript. Finally, we would like to thank Maria Finn for the beautiful and sensitive drawings she made for the book.

Maria Finn, *Unfinished #16*, pencil on paper, 29 × 42 cm, 2018. © Maria Finn.

Introduction

Prologue: Scaffolding

Copenhagen, May 2019

When my mother died, I bought a house that needed a new roof. Somehow crisis and construction, eruption and containment, loss and latitude often come as one package, I have found.

Henriette visited me, and I took her to the top of the scaffolding that surrounded the entire house at that point to inspect the construction work. Neither of us were very agile—Henriette seven months pregnant and I as always afraid of heights. We sat on the roof, perhaps six meters above ground, surrounded by scaffolding. We talked about physical scaffolds, emotional ones, and the theoretical scaffolding that we as academics build around the phenomena we want to explore in order to get at eye level with our objects of study.

We had been working together for some time on the changing connotations of transparency and invisibility today. We had received a grant to host a range of conferences; we had brought together scholars from around the globe and from different academic disciplines for discussion; we were editing books and special issues and cowriting a number of articles. Yet although we could see that our ideas had struck a chord, we had trouble finding ways to word exactly what we were looking at, as if the questions themselves needed support, were unsafe, and left us vulnerable. I do not recall if this was when the idea for this book first took shape. As I think back, the idea was definitely latently present that quiet afternoon atop my house stripped of its roof. Depending on one's perspective, it is an opportunity, a conundrum, or an impasse that the question of how to erect something is somehow silently present in sites of destruction.

Remains of scaffolds have been found in caves in southern France, where they are believed to have supported those who painted the walls with reindeer, aurochs, mammoths, and horses seventeen thousand years ago. In Hong Kong, scaffolds of bamboo with nylon string, up to 100 meters high, are not uncommon. In the European Union, the design and erection of scaffolding is

towertotower Copenhagen

Liked by **mariafinn_** and **1 other**

towertotower Wrapped in scaffolding #roof #scaffolding #framework #entrapment

regulated to ensure that "the purpose of a working scaffold is to provide a safe place of work with safe access suitable for the work being done."[1] According to this standard, safety should be independent of the materials of which the scaffold is made.

Yet the standard does not seem to account for theoretical scaffoldings constructed to support arguments and words. Although the scaffoldings used in construction work and even emotional scaffoldings are somehow held in check by the object (or subject) that they support, there is always a danger that theoretical scaffolds might be supporting something that is not really there—that, in fact, they are creating the gigantic object whose construction they support as they raise us inch by inch above the ground.

This chapter outlines a theoretical and methodological framework for the investigation carried out by the book you now hold in your hands. By moving from the Eiffel Tower to the Twin Towers and then to One World Trade Center and beyond—from one gigantic tower to the next—this book charts gigantism as a significant phenomenon of the present cultural moment and considers its ties to the past. The chapter asks *why* we should pay attention to gigantism today and suggests *how* we might go about it. But we also discuss the risk that we become part of the vanishing ontological borderline with which our study of gigantism engages—a borderline that we sense as an entrapment, an uncomfortably sticky position from which to work. So what kind of balancing act—on a wobbly scaffolding with no railings—do we need to perform if we are to tease out the workings of gigantism today?

—KV

A Tale of Two Cities, A Tale of Two (and More) Towers

> It was the best of times, it was the worst of times, it was the age of wisdom, it was the age of foolishness, it was the epoch of belief, it was the epoch of incredulity, it was the season of Light, it was the season of Darkness, it was the spring of hope, it was the winter of despair, we had everything before us, we had nothing before us, we were all going direct to Heaven, we were all going direct the other way—in short, the period was so far like the present period, that some of its noisiest authorities insisted on its being received, for good or for evil, in the superlative degree of comparison only.
>
> —Charles Dickens, *A Tale of Two Cities* (1859)[2]

The gigantic is everywhere. Not just physically, in the buildings, infrastructure, and booming cities that mark the grand feats of Western industrial culture—feats that Dickens evokes in this passage, written in the midst of the industrial revolution in the 1850s—but also in the excessiveness of the dichotomous concepts used to describe modern culture as Dickens arranges them in his text. Nearly a hundred years after Dickens wrote this passage, in a text published in 1950, the German philosopher Martin Heidegger notably wrote that gigantism is one of the most fundamental and fraught features of Western modernity.[3] Today, as anthropologist Anna Tsing notes, we are at a juncture where modern industrial culture seems to imprint the entire planet, and this means that the scale of the gigantic is an increasingly evoked thought-frame for describing the unsustainable nature of industrial culture and the instabilities to which it has given rise.[4] This situation calls on us to turn our attention to concepts of gigantism as well as the gigantic.

Originally a medical term describing excessive growth, the word *gigantism* is used in this book to describe material and immaterial structures that appear so big and all-encompassing that they have an overarching, ubiquitous quality to them—although they are never available to or experienced in the same way by everyone everywhere. In this book, the word *gigantism* describes the excessive ways in which the people of modern Western culture have built large structures and thereby brought about enormous unsustainabilities (often masked as progress), vast inequalities (often masked as universalities), and gigantic utopianisms (often masked as meaningful relationalities). These gigantisms concern not only the built and social world but also theoretical discourses. Gigantism as a conceptual scaffolding is therefore key to understanding wider implications of the gigantic infrastructures that surround us, at the same time as the word's excess reminds us of the impossibility of calling them out as singular entities—a difficulty that marks the uncomfortable position we build in this book, where we have found ourselves impacted by and implicated in the various forms of gigantism we explore and critique.

To us, gigantism today raises concerns in relation to ecological, material, cultural, affective, economic, and resource sustainability, and we focus here on two significant markers of gigantism in the early twenty-first century in architecture and digital culture. On the one hand, we discuss urban planning's continued race to build tall by erecting ever taller skyscrapers. On the other hand, we consider the encyclopedic ambitions of datafication and the pervasiveness of the networked digital infrastructures that span the globe today. These two forms of gigantism—the architectural stretch toward the sky and the digital enveloping of the globe—intersect in the figure of the skyscraper with a transmitting antenna on its roof: gigantic buildings that are nodal points in gigantic digital infrastructural networks.

A key aim of our book is to describe today's particular intermingling of architecture and digital culture as *latent gigantism*, a form of gigantism our analysis of One World Trade Center in New York City helps to unravel. The building was erected in 2014, next to the site where the Twin Towers of the World Trade Center had stood until the buildings were destroyed in the terrorist attack on September 11, 2001. At 1,776 feet (541 meters) and with a huge antenna, powerful transmitters, and satellite dishes mounted on it, One World Trade Center is the tallest building in the Western hemisphere. Although the size of the building sets it apart—as does the symbolic investment of the ground on which it stands—the building's architecture is remarkably unremarkable, resembling many other recent skyscrapers on Manhattan and elsewhere in the world, an unobtrusiveness that becomes apparent when you see the tower creeping in from behind the urban scene in Maria Finn's drawing at the start of this introduction. Moreover, in the massive data streams that permeate and envelop the building and travel from the antenna at the tip of the structure, another significant dimension of the building's gigantism becomes apparent: as a node in global information infrastructures, One World Trade Center appears as a strangely understated symbol of those digital infrastructures that both fracture and bind together people and places in neoliberal urban economies today.

In order to more fully comprehend the cultural role and meaning of an elusive yet gigantic structure such as One World Trade Center and the way it rises out of the cultural history of the twentieth century, we start this book on a different continent and in a different century. In 1889, when the Eiffel Tower was erected in Paris, it was the tallest tower the world had ever seen, and at the top of the tower there was a technical observatory and laboratory that was used for scientific experiments. When radio transmission was invented only a couple of years later, it did not take long before radio signals were sent from atop the tower.[5] In fact, the story goes that the Eiffel Tower escaped its planned demolition twenty years after the 1889 World's Fair only because of its newfound function as a radio transmitting station. Today, the Eiffel Tower is one of the most iconic examples in Western culture of a tall structure culminating in a transmitting antenna, and we argue that it embodies key aspects of gigantism's shifting connotations since the end of the nineteenth century.

This book is thus a tale of two towers in the sense that we focus on two paradigmatic tower sites in Paris and New York City, although the number of towers discussed is greater than two. It moves from tower to tower and thereby provides a partial account of different iterations of gigantism in architecture and digital culture since the late nineteenth century and up until today. It focuses individual chapters on iconic modern high-rise buildings in Paris and New York and moreover considers other tall towers around the globe, from Malmö in Sweden to Beijing in China. We start in chapter 1 by considering the Eiffel Tower, built in Paris in 1889, and we discuss changing interpretations of the tower across the twentieth and twenty-first centuries. In chapter 2, we consider what New York City must have looked like from the viewing platform atop the North tower of the World Trade Center built in 1970, and we discuss the Twin Towers' architecture, urban design, and destruction in the terrorist attack in 2001. In chapter 3, we engage with the architectural debates surrounding the erection of One World Trade Center on the Ground Zero site where the Twin Towers used to stand, and we report from our own recent visit to the building's multimedia observatory on the 100th to 102nd floors. In chapter 4, we discuss the conditions for civic culture in and around Ground Zero, the public space next to the building, and its relationship to contemporary digital culture. As we engage with the cultural, architectural, and media history of all these towers from their erection until today, we therefore also consider the changing

Introduction 7

and clashing understandings of gigantism the towers represent, not just as concrete structures but also as sites for the projection of cultural ideas and ideals.

As much as this book is about striking pieces of architecture, however, it is also about the interlaced digital networks these towers facilitate and in which they are embedded. We study the towers *as* media (as cultural artifacts and communication hubs) as well as how the towers feature *in* media (in radio, TV, news, social media, and well-known cultural-theoretical texts). We end in chapter 5 with a series of discussions of recent gigantisms across architecture (such as the CCTV building in Beijing) and digital culture (such as the notion of the tech giant). These discussions help us to consider, widen, and challenge Western ownership of gigantism and its association with modern industrial and cultural production. Finally, we return to New York City and the planned and ongoing urban planning projects to protect Manhattan from rising sea levels and turn Staten Island's Fresh Kills Landfill site into a park.

The chapters develop chronologically in terms of our study material, combining cultural-theoretical and architectural readings, and they may be read individually as additions to existing scholarship on the buildings and sites in question. Seen together, the argument they build concerns different forms of gigantism and their shifting significance in the period the book covers. The modes of discussion we evoke bring together theoretical reflections and critical engagements with canonical and contemporary cultural and architectural theory. Furthermore, we call on our own situated experiences and observations, in particular with regard to One World Trade Center.

In addition, we triangulate the discussions in the chapters with two other strands. One strand is a series of short prologues at the beginning of each chapter, including this introduction and an epilogue at the end of chapter 5. The prologues strike a different, more embodied and poetic tone than the traditional scholarly subject position, and they present alternative pathways into our argument. They reflect the difficult and productively incomplete fusion of horizons that the collaborative experience of writing this book has offered, and thereby gesture toward the slippery forms of commonality we discuss. Writing this book has brought into dialogue our singular and situated knowledge and experience and has provided us with a way of working through and beyond the idea of the individual academic author. In this ongoing dialogue, we have found a way of sharing, which has been crucial for us to approach the gigantic objects and phenomena we discuss in the book—phenomena that always implicate us as well as others and in which we ourselves are sometimes implicated in uncomfortable ways.[6] Yet we have also encountered the limits of that sharing, and the prologues and the epilogue, including the discussion of scaffolding in the prologue above, are places where we reflect on those limits.

The other strand we use to triangulate the discussions in the chapters is nonverbal and consists of a series of drawings by artist Maria Finn, which provide visual reflections on the paradoxes of appearance involved in our discussion of the towers and their gigantism. The drawings have been conceived by Finn in dialogue with our developing manuscript and have formed important vehicles for our thinking.

The iconic towers we approach in this book and the digital communication their antennas embody on a gigantic scale offer a looking glass through which to understand how different forms of gigantism have intertwined in architecture and digital culture since the late nineteenth century until today, where they reflect a number of ecological, economic, and cultural crises. This historical perspective is significant because it allows us to recognize the continuities and interdependencies between material and immaterial forms of gigantism. The towers' iconicity is no less significant: their overexposure in media and as tourist sites, which threatens to render them "too much" and hence unworthy of scholarly attention, is precisely what makes them key to a study of gigantism. Our task in this introduction is to build what we in the prologue have named the theoretical scaffolding of the book by way of establishing the notion of gigantism, the different iterations of this term that govern our reading of particular sites, as well as a series of reflections on the terminology and methodology we employ. Readers more interested in the sites than the theoretical underpinnings may therefore prefer to jump directly to chapters 1 through 5.

Gigantism

The root of the word *gigantism* is the Greek word *gígas* meaning "giant," a human of very great size. It also refers to *gígantes*, savage and monstrous creatures in Greek mythology that were the offspring of Gaia, goddess of the earth, and Uranus, the sky, and which personified destructive natural forces. The adjective *gigantic*, then, meaning "overly large," gives rise to the word *gigantism*, which has been used in medical contexts since the mid-nineteenth century to describe a condition of abnormal overgrowth.[7] The suffix *ism* can designate a movement or distinctive doctrine (for example, in art or philosophy). Our use of it here intends to conceptualize gigantism as a form of excess, a diagnosis attached to very large things or phenomena. Moving from tower to tower in this book, we identify the gigantism that emerges when unsustainability is subsumed as a marker of progress, for example, through architectural form experiments on a gigantic scale in various historical examples of trying to build one of the largest structures in the world. We also consider the way inequalities can trade as universalities, for example, in the seemingly neutral space created by vast digital infrastructures. And we discuss what happens in key contemporary academic discourses when flattened relational utopias turn cruel or hide in even the most hopeful, daring, committed, and evocative countervoices to some of the gigantic normative narratives of modern Western culture.

Gigantism, we argue, is a way to engage central dilemmas and contradictions that we have inherited from the modern world. As anthropologist Anna Tsing has noted: "We learned to know the modern by its ability to scale up." But it has left us with two problems: "first, expandability has gotten out of control. Second, scalability has left ruins in its wake. Nonscalable effects that once could be swept under the rug have come to haunt us all."[8] Tsing argues that there is an erroneous belief today that everything is scalable, a belief emphasized in digital technology as much as in business culture. This belief in scalability has given rise to some of Western modernity's wildest and most unsustainable exploitations of people and of the earth's ecosystems. From a position in environmental humanities, Rob Nixon underscores this critique when he writes that large-scale infrastructural engineering projects such as megadams—or this book's skyscrapers—are products of modern culture's desire to mark ascent and greatness, often at great economic, social, and ecological cost, not least in postcolonial and developing-country contexts. He shows how vast unsustainabilities can hide behind gigantic infrastructural and architectural structures that are built as markers of progress, progression, and growth. Nixon even calls this expansionist desire "a disease . . . of developmental gigantism,"[9] a slow violence, and thereby emphasizes how vast inequalities can trade as universal goods in apparently neutral and benign infrastructural frameworks.

Characters in Canadian writer Margaret Atwood's 2003 postapocalyptic speculative novel *Oryx and Crake* articulate this impasse when they state that the agricultural foundation of the whole of human civilization has been subsumed into one destructive narrative with a force as powerful and mythically charged as that of the giants of Greek mythology:

> It had been game over once agriculture was invented, six or seven thousand years ago. After that, the human experiment was doomed, first to gigantism due to a maxed out food supply, and then to extinction, once all the available nutrients had been hoovered up. . . . Maybe there weren't any solutions. Human society, they claimed, was a sort of monster, its main by-products being corpses and rubble. It never learned, it made the same cretinous mistakes over and over, trading short-term gain for long-term pain. It was like a giant slug eating its way relentlessly through all the other bio forms on the planet, grinding up life on earth and shitting it out the backside in the form of pieces of manufactured and soon-to-be obsolete plastic junk.[10]

The mollusk's excrement in this extract is a metaphor for the dumb sameness of consumer society, a critique that has often accompanied modern popular culture (in the tradition of Theodor Adorno and Max Horkheimer, for example) and that is politicized by Atwood in the way the mollusk trades lives and real bodies for dead, devoured objects that have no deeper meaning. The giant slug pretends to move forward yet in fact moves toward an abyss. It embodies gigantism as a product of modern culture that requires substantial critique, but at the same time, it also embodies a position from which critique itself can slide into dystopian (as much as utopian) understandings.

Therefore, the idea of sameness that Atwood metaphorizes by the mollusk's excrement marks a key characteristic of vocabularies that conceptualize the gigantic—a flattening or ontological slippage, of which we see different variants across the chapters of this book. If we take the neutrality of all-encompassing terms such as *gigantism* for granted, we ignore the fact that even the most gigantic phenomena are also embedded, embodied, and situated. And we may easily forget that they come with particular affordances that are gendered and racialized, for example, and that often privilege the already privileged, loquacious, or able-bodied. With feminist scholar Donna Haraway, we therefore want to stress that "staying with the trouble requires making oddkin; that is, we require each other in unexpected collaborations and combinations, in hot compost piles. We become-with each other or not at all."[11] It is within this intricate compost pile that we find ourselves—the same but different and dependent—and this requires careful consideration of our relationships with nonhuman and material cultures, and calls for careful attention to dependencies and differences.

However, as Haraway often acknowledges, a too narrow consideration of relationships runs the risk of creating utopian positions that have implications of gigantism of their own.[12] As we show in our grappling with both modern and postmodern epistemologies as well as with relational and utopian thinking, they can come to constitute gigantisms in their own right (not least in the echo chambers of academic theory). Indeed, hopes invested in relational thinking to overcome the binaries of normative modern concepts can turn *cruel* in the sense used by Lauren Berlant when she speaks of "cruel optimism" as descriptive of a relation where that which we desire in fact forms an obstacle to obtaining that goal—for instance a particular fantasy of the good life, food or a particular kind of love.[13] We contend that there is a cruel form of optimism at work in the belief that "bigger is better," and that our attachment to and dependency on large structures may in fact also pose a threat to the life that takes place in and around these structures. Similarly, there is a cruel optimism at work in the fact that our relationships to gigantic structures can be theorized in ways that focus on their universal and inclusive potential at the same time as they are not equally accessible to all. Examples include the buildings, digital infrastructures, and theoretical concepts we discuss in this book. In this light, Atwood's characterization of human culture as a (perhaps unwittingly) monstrous yet relentless mollusk can be read as a metaphor for the paradox of what we call *latent gigantism*. Latent gigantism describes the puzzling situation whereby a phenomenon is so overwhelmingly large that it simultaneously withdraws into the background by virtue of its excessive scale, and in this way naturalizes the optimism of gigantic projections so that the way they might be "cruel" is left unaccounted for.

This slide is precisely what is at stake in the notion of the hyperobject as proposed by philosopher Timothy Morton in order to describe phenomena such as digital networks or weather systems that are so large that they cannot easily be grasped by humans. The hyperobject as Morton develops it calls attention to the kinds of gigantic objects produced by or influenced by humans, particularly those that are not easy or even possible to delimit clearly. Morton moreover theorizes the experience of local manifestations of these large phenomena, such as the interfaces of digital communication platforms or the effects of climate change, although those local manifestations are not the hyperobject itself.[14] However, as noted by cultural historians Michael Tavel Clarke and David Wittenberg, while Morton embraces the hyperobject as all-encompassing and ambient, his concept is also reminiscent of the Kantian sublime. The excessiveness of scale of the hyperobject fixates the rational human subject, which—even while fearfully experiencing the limits of its own cognitive abilities—may seek to place hyperobjects at the "right" distance so that they can be appreciated for their aesthetic qualities.[15] The idea of the hyperobject therefore exemplifies the ease with which theory can slide into gigantic flattenings of difference and—unwillingly, perhaps—fixate modern modes of subjectivity in our effort to comprehend gigantic phenomena that in fact destabilize such modes. The hyperobject therefore illustrates this book's central ambition and difficulty—to address not only the gigantism of towering structures and digital infrastructures but also the gigantism of the theoretical perspectives through which we approach them.

For something to be identifiable as gigantism, there needs to be something *more* at work than simply a large scale. Heidegger notes this elusive quality: "[As] soon as the gigantic in planning and calculating and adjusting and making secure shifts over out of the quantitative and becomes a special quality, then what is gigantic, and what can seemingly always be calculated completely, becomes, precisely through this, incalculable."[16] In describing the gigantic as incalculable, Heidegger points to a shift or slide away from largeness in quantitative terms and toward the gigantic as a special quality that takes on a significance of its own and is therefore difficult, if not downright impossible, to measure.

The gigantism of the towers we study in this book does not lie in the feet or meters they reach into the sky, in the tons of concrete or steel that give them shape, or in the amount of wire used to hold them together—although these quantities can be used to assess their sustainability.[17] Rather, their gigantism emanates from the bodily and affective impossibility of any sensation of scale when one encounters these structures. Likewise, the gigantism of digital infrastructures does not lie in the number of likes or shares that an image receives on Instagram or Facebook, in the miles the information travels as it pulses through wired or wireless connections, or in the terabytes of data used for big data analytics—although these quantifications are central to how digital publics are organized and managed in groupings of "people like you."[18] Gigantism arises from the unpredictability of the range and impact of these information structures and from the incomprehensibility of the effects of the digital traces that users of digital infrastructures leave behind.

From their different perspectives in phenomenology, posthumanism, and feminism, writers as different as Martin Heidegger, Anna Lowenhaupt Tsing, and Donna Haraway mark out the need to investigate gigantism's ties to modernity. When approaching gigantism today, as we endeavor to do in this book, our task is therefore to nuance our understanding of the continuities and differences between contemporary forms of gigantism and Western modernity's forms of gigantism. What unsustainabilities, inequalities, flattenings, and utopianisms do these concepts carry, and what is their impact? What happens to our understanding of architecture, cities, media, and the position of the modern human subject? In other words, what happens when Western modernity's forms of gigantism are increasingly unsettled in a wider global perspective?

Vertical and Horizontal Gigantism

As architectural constructions, the towers considered in this book allow humans to ascend them, and they are equipped with viewing platforms or panoramic windows for people to look out of. In the transmitting towers, we see forms of gigantism that operate not just vertically (that is, the tower reaching for the sky) but also horizontally (the global span of the distributed communication networks that tall buildings facilitate). Their antennas enable the reception and transmission of information in a way that bestows on the towers a kind sensory capability beyond that of the human sensorium. The antennas of these transmitting towers are therefore more akin to the workings of an antenna on an insect than to the monodirectionality of a radio mast. Architectural theorist Mark Wigley aptly reminds us how this "antenna ecology" profoundly interlinks human, architectural, and digital cultures to an extent that makes it necessary to rethink the relationships between humans, machines, and architecture.[19] As he states:

> Antennas are in the smartphones in every pocket, they are in every room, building, street, vehicle, fish, bird, drone and product in the supermarket. We swim in a completely interconnected ecology of antennas, from the depths of our intestinal

tracts to interstellar space. What has happened to architecture as a result? Antennas destabilize architecture. They redesign it. Figuratively speaking an antenna can dwarf any building to which it is attached, providing levels of communication and shelter far beyond the architect's most ambitious imagination. On the other hand, an antenna can supercharge the most modest building. A humble shed can be launched into interplanetary exchanges with the addition of a thin wire connecting the building to an invisible world of signals.[20]

The skyscraper with the transmitting antenna on its roof is exemplary for capturing a contemporary form of digitally wired gigantism with a strong modern heritage. As a structure, it points to all the descriptive dichotomies of modernity that Dickens evokes in our epigraph from the novel *A Tale of Two Cities*. The upward stretch of the high-rise tower implies the normative connotations of heaven and hell, light and darkness, hope and despair that Dickens mentions. The extremism implied in Dickens's remark that the times of his novel were received "in the superlative degree of comparison only" reads strongly today at a time when political frontiers are being drawn sharply in the rhetoric of superlative degrees. The history of the skyscraper and its contemporary persistence as an architectural form that appears simultaneously to be cutting edge *and* tediously dated can help us to understand the uneasy sense of anachronism that can come from the rhetoric of the ultimate vertical frontier.

The assumption that "bigger is better" remains the case despite the predictions of the end of the high-rise following the terrorist attacks on September 11, 2001, in New York City and in spite of the high-rise's position as both a potential terrorist target and the symbol of an urban planning paradigm whose financial and environmental sustainability is increasingly dubious—a position that challenges the normative evaluation of "up" as a common good. Moreover, since the invention of long-distance communication technologies—not least the invention of radio transmission around the time the Eiffel Tower was built—the vertical axis of the tower reaching into the sky has been crisscrossed with wired and wireless signals. Today, connotations of progress and conquest are implied in the idea of emerging digital technologies as new frontiers, just as the history of the internet is laden with the terminology of the gigantic, pertaining to speed as much as size (such as the worldwide web and the information superhighway to mention just a few).[21] Gigantism also continues to thrive in digital culture, as seen, for example, in the way machine learning algorithms promise to encompass and operationalize more and more of the world's data, while also accentuating preexisting biases and uncertainties.[22]

Although the twentieth century has now slipped into the new millennium, the high-rise tower with a transmitting antenna on its roof remains a symbol of power. It shoots up across the globe in diverse political and social contexts but with a remarkably uniform symbolic investment,[23] the implications of which this book seeks to unravel. Significantly, the gigantism of these towers today stems as much from their function as nodes in the information infrastructure that envelops us as from their towering physical dimensions: their gigantism is horizontal as much as vertical, and it is marked by messy overlaps between the two. On the one hand, we have to do with a symbolic language that emphasizes linearity, progress, and progression as an overcoming of humans' earthbound existence. We address this as *vertical gigantism*. On the other hand, it is part of a digitally enhanced present that is highly distributed both temporally and spatially, carrying ambitions to encompass not only the globe but also the past, present, and future by means of digital technologies' emphasis on temporal overlay, simultaneity, and synchronicity. We address this as *horizontal gigantism*. In this book, we tease out the ways in which different forms of gigantism make themselves present in the architectural sites we visit, and we show how they are mutually dependent and intertwined. We do not regard vertical and horizontal gigantism as opposites; rather, we see them as different registers that articulate similar ambitions. As nodal points in a landscape where horizontal and vertical gigantism intersect, however, high-rise towers are sites from which we can study the current forms and effects of gigantism in architecture and digital culture.

The intermingling of vertical and horizontal gigantism in high-rise towers with antennas on their roofs asks for a methodology that borrows from different disciplinary traditions. As architectural scholars, we can start our investigation by climbing the gigantic structures we study—walking or taking an elevator to the top of the tallest built artifacts in our upward-expanding cities. As scholars of the digital fabric of the contemporary city, we can likewise take a ride outward in gigantic digital networked structures, through the various interfaces we operate, to get a glimpse of their expansiveness as infrastructural systems dependent on digital structures of code. As cultural researchers, we can trace, theorize, and critically examine the trajectory of the emergence of these infrastructures, as well as their reflections in the images and narratives that pervade our cultural imagination—ranging from memes and viral YouTube clips to film, art, and literature. And we can interpret the ways the towers make themselves present in the gigantic and self-reproductive digital infrastructures that mediate, produce, and archive contemporary culture. Moreover, we can use our bodies and emotional registers to gage the towers' architectural and mediatic effects and affects.

In doing all this, we have found that the architectural and digitized realms of gigantism—realms that have material and mediated aspects of their own—are in many ways two sides of the same coin. They both link up with modern culture's fundamental drives for growth and progression. They do so in a way that connects with dualistic categories of opposites where extreme scales of largeness (or, conversely, extreme scales of the very smallest, as in technological downscaling) become markers of grandeur, significance, success, and beauty on their own terms and are contrasted with what is less spectacular in scale and hence classified as insignificant, failed, or ugly. Inspired by cultural theorist Sianne Ngai's discussion of minor affects and aesthetic categories[24] and by literary scholar Susan Stewart's discussion of the miniature and the gigantic,[25] we therefore engage with towers that are at the same time overexposed (for example, as tourist symbols) *and* curiously minor in academic discourse.

In the individual chapters of this book, we discuss examples of iconic high-rise towers with and against a number of iconic thinkers whose writings have helped to define the status of such buildings and their urban, political, and discursive contexts, including Roland Barthes, Jean Baudrillard, Walter Benjamin, Michel de Certeau, Donna Haraway, Martin Heidegger, Bruno Latour, and Slavoj Žižek. These writers have produced parts of the central corpus of twentieth- and early twenty-first-century cultural theory, and are notorious for offering perspectives that fracture dualistic thinking or the very idea of Western culture's power and centrality, including the positive connotations of growth, progress, and gigantism. Yet their towering position as cultural authorities compels us to examine critically and historicize their writings in the same way that we critically examine and historicize iconic pieces of architecture. Therefore, we tease out the continued relevance of the commentaries they wrote at key moments in the reception of the towers in question, but at the same time we also critically reflect on their thinking. To do so, we draw on a diverse group of contemporary critical cultural, media, and architectural theorists—among them are Sara Ahmed, Lauren Berlant, Mario Carpo, Wendy Hui Kyong Chun, Saidiya Hartman, Ian Hodder, Shannon Mattern, Rob Nixon, Zoë Sofoulis, Anna Tsing, and Joanna Zylinska. Along with the many other writers we implicitly or explicitly draw on in this book, they have all influenced our thinking in important ways.

Bringing these different theoretical traditions together is a challenge. Throughout our writing, we have found ourselves confronted with the descriptive dichotomies of modern culture—categories such as light and darkness, eruption and stasis, verticality and horizontality, surface and depth, mind and matter, bodily and virtual presence. Although we are indebted to various theoretical attempts to break down such dichotomies, we also recognize that attempts to move beyond them sometimes entail ontological slippages that conflate categories such as the human, the technological, and the material. Sometimes attempts to challenge dichotomous thinking come to mimic this conflation, as you will see in more detail throughout the book.[26] We discuss evidence of ontological slippage in the material we study, in the theoretical apparatus we employ, as well as in various constellations of relationships between the two. We are concerned by the parallelism between our theoretical and empirical material, and we describe its effects as a form of entrapment (building on archeologist Ian Hodder's use of the term)[27] in order to comprehend how our own vocabulary is implicated in what we are trying to describe. We confront this situation by attempting to historicize the different and sometimes incongruous thinking and concepts we draw on, as well as by pointing out our own indebtedness to previous theoretical wrestles with the gigantic dichotomies of modern culture. Our challenge to bring different theoretical traditions together is therefore connected to questions about how the relationships between humans, technology, and materialities play out at the sites we study and how categories of past, present, and future seep through them. It implies a reposing of other and perhaps more familiar questions: can we challenge modern epistemologies while still drawing on them? Is it possible to analyze gigantism's all-encompassing character at the conceptual level without reproducing this gigantism conceptually?

Gigantic Dichotomies, Containers, and Leaks

To exemplify these notions of parallelism and entrapment and their relation to our readings of gigantic towers, let us turn to gender categories that we have found (unsurprisingly, perhaps) even in the most unexpected of places when dealing with interpretations of very tall rectilinear buildings. When we explore the gigantism of tall towers as blatantly phallic structures, our critique is informed by the concerns raised in contemporary feminist and queer critiques of large-scale infrastructures of modern culture, such as the work of cultural theorists Sara Ahmed and Zoë Sofoulis.[28] Unsurprisingly, the rhetoric of scale expressed in the big-small dichotomy can be politicized in relation to gigantic tower buildings, and when we move from tower to tower in this book, we see glaring attempts to assert dominance through the bigger, better, and taller. At the same time as the towers we study in this book protrude into the world, however, they are also containers—of people, wealth, power, and everyday urban life. By reading them not only as rectilinear forms but also through the metaphorics of the container—in line with the ways feminist theories have engaged with the latter—we tease out the significance of the gigantism at work in a way that cuts across gendered dualisms, elucidating how power operates through various gendered categories, not only through masculine typologies.

American novelist Ursula Le Guin links gender, gigantism, and power when she writes that the history of *man*kind is written from the perspective of the man as hunter, where history is tracked as a series of advancements, such as the control of fire and the invention of the spear. She argues that these inventions in and of themselves would have been insignificant if there had been no "carrier bag" or container to bring back the prey, and in this way she points to the significance of the work of the gatherer, historically a position ascribed to women.[29] Zoë Sofoulis also combines questions of the container, gender, and power relations when she speaks of container technologies.[30] Building on the classic work of American historian Lewis Mumford as well as on philosopher Martin Heidegger, she reassesses container technologies as utilities that contain, store, preserve, and protect things and that often connote femininity, as opposed to tools such as hammers or spears that extend the (male) body. Container technologies are techniques of the unobtrusive, the contextual, and the environmental and are productive simply by being there, holding things together. They are technologies that Mumford's 1934 study categorizes as utensils (baskets, pots), apparatuses (dye vats, brick kilns), utilities (reservoirs, aqueducts, roads, buildings), and modern power infrastructures (railroad tracks, electricity lines).

Sofoulis emphasizes the necessity of correcting phallic biases and confronting dichotomous thinking in interpretations of technology and traditional Western notions of space as passive and feminine—particularly in light of so-called smart technologies that operate at the intersections between entity and environment, tool and container. When she gives examples of such container technologies, she correlates the idea of the phallic skyscraper with questions of digital technology in a way that aligns with our argument in this book: "Examples here include the skyscraper, so obviously phallic but from the inside a 'womb with a view'; the car, advertised in terms that emphasize on the one hand its phallic/excremental 'grunt,' and on the other its womby comfort and storage space; and the computer, which is basically a storage technology for data, yet which has often been represented as a kind of flying vehicle, even before widespread networking allowed internet 'surfing.'"[31]

These qualities of container technologies allow us to unlock surprisingly rich cultural imaginaries that cling to the towers of this book. As containers, these towers can effectively hold in place, but also *leak* information about what is inside them, to invoke a term that is currently used primarily in relation to digital culture but whose strong material connotations make it apt for the realm of architecture.[32] As Sofoulis evocatively argues: "Not all containers are designed to be impermeable or like the jug capable of outpouring: some are for slow leakage, some for soaking up drips, others for what we hope will be permanent containing. An extended analysis of containers would have therefore to examine 'incontinence'—various deliberate (as in a colander or coffee filter), catastrophic (like the *Titanic* or Chernobyl), or merely embarrassing(!) failures of containment."[33] The notion of leaks works on two levels throughout this book. On the one hand, we observe and identify leaky qualities in the material we engage with—from the Hudson River leaking into the foundations under Ground Zero to the information leaks of hacking tools. On the other hand, and on a more conceptual level, leaks are ways of engaging with differences and dependencies that cannot be fully understood through the vocabulary of ontological flattening.

We moreover read the leak in light of "stickiness" as conceptualized by Sara Ahmed's affect theory. When an object becomes "sticky," it is saturated by affect, and stickiness is thus *"an effect of the histories of contact between bodies, objects, and signs,"* Ahmed writes.[34] Stickiness, as we use it here, thus marks a difference but also a dependency that lingers, sometimes latently. Leaks have a sticky quality to them, which makes them rich and complicated metaphorical vehicles for understanding dependencies. In this book, we offer an analysis of the way gigantic towers are phallic superstructures from which wireless signals travel far into an abstracted informational space, at the same time as they are also containers of people and information that provide concrete viewing platforms onto the historical, cultural, social, and material conditions out of which they were born. We search for ways not so much to dismantle the gigantic dichotomies of modern culture but rather to recognize, theorize, and historicize the flattening that occurs in doing so and to understand the difficult possibilities for knowledge that arise in this situation. The double perspective on towers as erect structures as well as containers allows us to think about the points where the gigantism that marks the towers breaks apart or allows us to pay attention to the dependencies and differences at work—in other words, where they leak and thereby reveal more complex temporalities and histories that "stick" with us.

Tower to Tower

This book moves from tower to tower: from the Eiffel Tower (erected on a whim for a World's Fair in Paris as a manifestation of industrial and colonial power) to the World Trade Center (built to inaugurate a new age of global high finance in an attempt to establish New York City as the center of the Western world) and then to One World Trade Center (built to replace to the destroyed World Trade Center towers and currently the tallest building in the Western hemisphere). The iconic towers we focus on in this book have been selected precisely because they are blatantly gigantic relics of a Western industrial culture steeped in colonial mindsets.[35] Inherent in these incarnations of gigantism, we suggest, is a Western nostalgia for a grandness that seems massively irrelevant, inadequate, or even harmful in the face of ecological crises. Yet it is a nostalgia that defiantly lingers on in many places, including politics, as in the current US President Donald Trump's notorious slogan "Make America great again." This quest to reinstall greatness in the United States of America not only suggests that this country lacks greatness but also that it once had it. This crisis of "greatness," which presents the other side of the coin of this statement, is thus marked by a narrative of disappointment. In this light, it might be significant that the buildings that lead the race for tallness today are found elsewhere than the Western world. The gigantism of skyscrapers in Western cities may have a tired, outdated, or even anachronistic air, but markers of gigantism worldwide are no less prevalent than before, even though—as we show in this book—they may differ from the rectilinearity of the skyscraper. Moreover, as Western culture's crises multiply in relation to material, ecological, economic, and social factors, these buildings make it apparent that gigantism reflects long, complicated, global and colonial histories.

Needless to say, Paris and New York City occupy central positions in the Western cultural imaginary as world cities of the nineteenth and twentieth centuries. Although the majority of the discussion in this book centers on Paris and New York City, we also discuss other urban sites and situations, as well as the gigantism inherent in ideas of global or Western sameness—"the global" and "the West" themselves being concepts that need serious questioning. The iconography of Paris and New York continues to reverberate across the globe in new urban structures in the twenty-first century, but their centrality to the global and political economy seems to be diminishing and arguably has been doing so for a considerable time. By focusing on both a European and an American example, we emphasize gigantism not just from a US perspective of "bigger and better,"[36] as gigantism is often described, but in a broader context of Western culture at the same time as we challenge these narratives and see them as part of the gigantism we explore.

The cities and sites we investigate have shared histories that make them stick to each other in ways that traverse space and time. Maria Finn's drawing on page x shows a view of the Eiffel Tower from a bridge in Paris, where we see the original French model of the Statue of Liberty, and as this view indicates, the historical events of the French Revolution and the American narrative of independence and freedom are intricately intertwined. The statue was a gift from the people of France to the people of the United States, and it was erected in 1886, just three years before the Eiffel Tower was built. Although the Statue of Liberty was designed by French sculptor Frédéric Auguste Bartholdi, its metal framework was built by none other than Gustave Eiffel, the engineer behind the eponymous tower.[37] Such historical collusions also account for other sites we investigate. For example, as we discuss in chapter 5, the CCTV building in Beijing designed by Dutch architect Rem Koolhaas in the early 2000s, around the same time as the design competition to rebuild Ground Zero, reveals itself as a consciously twisted formal echo of the Twin Towers' architecture. This relationship both complicates and extends narratives of "the West" and "the other" on a gigantic scale.

However, these historical links are not the only ones that connect the buildings, sites, and cities in question. In recent years, many of the gigantic landmark towers we discuss in this book have been associated with terrorist attacks in one way or another. Although terrorist attacks in cities are not the subject of this book, the terror linked to the towers in question and its mediatized reverberations have added important chapters to the towers' histories. We read the attacks in New York City in 2001 and Paris in 2015 as examples of how disruptive events can provoke profound, large-scale expressions of cultural concordance that carry connotations of gigantism, not least in the way that gigantism glosses some events as more grievable than others.

It is striking how situations of crisis illuminate changing understandings of cultural concordance and commonality and how interrelations between architecture and digital infrastructures recalibrate notions of civism and publicness in cities. The idea that disruptive events can provoke expressions of solidarity on social media from people located in places that are geographically far removed from the events themselves taps into the utopian idea of a digitally connected, homogeneous global village that is wholly inclusive.[38] However, it thereby reveals the limitations of the idea of global resonance and of the very concept of "the global." The presence of inherent biases and blind spots in these global responses to local events reveals that a certain abstraction is taking place. This abstraction is part of what we address as a flattening, an erasure of difference: the underlying narrative that digital infrastructures are neutral glosses over gendered and racial implications, for example.[39] Insofar as such flattening bears the legacy of colonialist efforts to install global regimenting infrastructures, it hides behind nostalgic notions of gigantism that frame colonialist infrastructures as normative. As we argue, the flattening of difference that gigantism fosters is being reinforced by theoretical and analytical perspectives that are also premised on flattening. As we move, in the chapters of this book, from the birth of the Eiffel Tower to the present day, we reveal that efforts to take solace in potent architectural forms and in the temporalities of linearity and progress they imply are increasingly emptied of cultural relevance. But we also show that these temporalities are becoming blurred by more ambiguous and distributed regimes of cultural signification that attach to a more latent form of gigantism.

Latent Gigantism

Media studies scholar Shannon Mattern reminds us that the relationship between architecture and digital culture, between cities and data, has a millennia-long history in which architecture and digital culture (or clay and code, dirt and data, as she writes) intermingle.[40] We may therefore regard cities as having always been mediated, she argues. Like Mattern, in this book we also go back in historical time in order to understand current forms of intermingling between architectural and digital cultures. Although the period we study is centennial rather than millennial—stretching from 1889, when the Eiffel Tower was erected, until today—this temporal demarcation should not be taken as a desire to think in historicist categories of ages or epochs. Rather, it is a framework that allows us to zoom in and consider in detail the complex temporalities at work in concrete sites and across the histories of individual buildings—all the while emphasizing architecture's role in establishing rich settings for urban culture, as Peter Carl argues from architectural philosophy.[41] The chapter-long readings allow a high level of specificity at the same time as the over 125 years that our studies span allow us to identify and reevaluate dominant temporal narratives (such as history's linear progression) and the spatial metaphors of origins, progress, ruptures, and peaks that accompany them.

Martin Heidegger, in his discussion of gigantism, argues that "each historical age is not only great in a distinctive way in contrast to others; it also has, in each instance, its own concept of greatness."[42] This remark links up the concept of gigantism with ideas of dominance, narratives about history as a succession of "ages," and historicism's narratives of history's progression. Although we situate different understandings of gigantism in different historical contexts and we propose three temporal categories to structure our readings of the historical period we span—*linear gigantism, semantic gigantism,* and *latent gigantism*—we are primarily concerned with the continuities between them. Moreover, we regard the narrative of history's linear progression as a form of gigantism in itself—one that is intimately connected to the way the motif of the potent tall building taps into central modern narratives of human accomplishment and of hope in progression and newness. The category of *linearity* is teleological, describing the gigantic as having a direction toward the future, but the multivalency of meaning implied in the term *semantic gigantism* concerns primarily language and visual form. The third category, *latency*, is ontological and concerns a form of gigantism that is latently present as an ontological and temporal flattening; it is perhaps the most gigantic of all. While linearity finds its clearest articulation (perhaps unsurprisingly) in the Eiffel Tower, the Twin Towers of the World Trade Center allow us to engage more thoroughly with semantic gigantism. Against the backdrop of these discussions, we see latent gigantism as most evident in One World Trade Center and in the wider examples we consider in chapter 5. We discuss linear, semantic, and latent gigantism in relation to the various towers we investigate. The three categories never quite replace one another but can be said to be present side by side at the sites on which we focus. We note the effects of these temporal categories reflected in our empirical material but also in the cultural theory on which we draw.

Linear gigantism is linked to what German historian Reinhart Koselleck defines as "historical time" and identifies as typical of modernity from the mid-nineteenth century onward, given the connotations of linearity, progression, and even progress that come with the idea of historical time.[43] Yet as we show, even when it comes to some of the tallest buildings of Western culture in the twentieth and twenty-first centuries, this sense of time as linear progression is in different ways (and arguably increasingly) overlaid by other temporal categories. Although linear gigantism emphasizes an excessive linearity and the forward march of history—prominent tropes in late nineteenth-century Western culture—it also reflects both vertical and horizontal forms of gigantism. But linear gigantism is not the only form of gigantism we see in the towers, nor is linear progression the only temporal category marking them. We also see more explicitly networked forms, not least on the semantic level, where the towers' architectural language and later media reception are marked by a potential for the endless repetition and expansion of meaning as a form of semantic gigantism. Semantic gigantism represents an excessive self-reflexive articulation of processes of meaning-making as significatory play that has been prominent, for example, in poststructuralist continental theory and postmodern architecture since the late 1960s.

Finally, latent gigantism, a concept we introduce in this book, reflects gigantism across architecture and digital culture in the early twenty-first century. Latent gigantism emphasizes that the singular present moment is knotted up with immediate and more distant pasts, as well as with projections into the future. The term designates the paradoxical situation where gigantism has become so familiar and taken for granted that it resides in the peripheral, ambient, calm, contextual, or environmental. We want to capture this with the notion of latency, which simultaneously connotes invisibility, silence, and delay, as well as overload, noise, bias, and uncertainty.[44] The term *latency* has a rich and diverse set of cultural-theoretical implications. Etymologically, *latent* originally meant "concealed" or "secret." It comes from the Latin *latentem* (nominative *latens*), the present participle of *latere* (meaning "to lie hidden, lurk, be concealed"), which is also related to the Greek *lethe* ("forgetfulness, oblivion") and *lethargos* ("forgetful"). In their large-scale and grandiose iconicity, the gigantic towers in this book simultaneously blend into their surroundings and draw attention to themselves, as do the digital infrastructures they feed and facilitate. They protrude into the world through sheer size, yet at the same time they are so familiar to our gaze that we hardly see them. Latent gigantism thus accentuates the paradox of gigantism that we explore in this book by conflating visibility and invisibility and by indicating not only an ontological but also a temporal flattening.

Across the readings of the towers in this book, we show how linear and semantic forms of gigantism and the temporalities to which they belong have increasingly been challenged by a different temporal metaphor that also has gigantic implications for the relationship between architecture and digital culture—the metaphor of a broadening present that conflates past and present into an extended "now." As cultural historian Hans Ulrich Gumbrecht writes: "Between the pasts that engulf us and the menacing future, the present has turned into a dimension of expanding simultaneities. All the pasts of recent memory form part of this spreading present."[45] This is accentuated by digital media's capacity for storing information, a process that can be portrayed as allowing the past to linger latently in our midst, mingling with the future as a perpetual part of the present and thereby determining the knowledge we can build of either. For example, certain risks can be predicted and preempted through the application of network analytics to the digital traces we leave behind, but this also reflects back on the way many of our mental faculties, from wayfinding to mnemonic capacities, are increasingly interlinked with information processing and data storage.[46]

Moreover, we contend that the broad present should be seen in relation to another gigantism that Peter Engelke and J. R. McNeill have called the "great acceleration." This concept describes the increasing speed with which humans impacted the global environment in the second half of the twentieth century and is interlinked with the idea of the Anthropocene.[47] In a parallel gigantic movement of its own, this extended temporal condition contributes to creating a slippage not only between mind and matter, locality and telepresence, and other fundamental dichotomies of modern culture but also between past and future. As a concept, the Anthropocene periodizes human culture's effects on the earth's ecosystems. This concept was proposed in 2002 by geologist and Nobel Prize winner Paul Crutzen as a way of conceptualizing humans' increased impact on the global environment.[48] The notion is widely used, but it has also been criticized from positions as different as environmental humanities and critical race theory—for example, by Andreas Malm and Kathryn Yussof, respectively. Both Malm and Yussof regard the notion of the Anthropocene as too tightly connected to an understanding of the human as a modern, Western, industrial, white, and largely male project—that is, it equates the human (*anthropos*) with the small group of people who initiated the industrial revolution with all its consequences.[49]

Bearing these different and often contradictory approaches and critiques in mind, we can see the Anthropocene scale as characterizing a type of gigantism similar to the idea of the hyperobject. The discourse around the Anthropocene resonates with the flattening alluded to in Morton's hyperobject, but also signifies the wailing cry of a historically instituted sense of subjectivity that belongs particularly to the modern Western world. It therefore serves to question the modern Western subject's sense of monopoly of self-reflective thinking or even reflective thinking as such, as emphasized in posthuman scholarship by scholars like Donna Haraway and N. Katherine Hayles.[50] If notions such as the broad present, the Anthropocene, and the great acceleration—which fuse temporal and spatial categories into a plane of projection for both optimistic and dystopian visions of the human condition—have a bearing on how scholarly or politically informed critique can take place, we are left with a situation where these notions are intrinsically stuck to the transformations they denote.

Latent gigantism points to an ambient form of gigantism where time coagulates and where the distinctions between human and material hover on a threshold. In contemporary cultural theory, aspects of this condition are conceptualized through a variety of terminologies that describe relational structures, using words such as *hybrid*, *assemblage*, or *entanglement* to do so—terms whose use can eradicate differences and thereby entail an ontological slippage. But it is continuities, differences and dependencies to which we wish to pay attention here and for which we have found the available theoretical vocabulary often falls short. Our recurrent focus on *leaks* in this book offers a way of grasping fine and often uneven differentiations, and we explore instances where these differences and different temporalities collide or lead to difficult, messy, and unsolvable entrapments—for example, when we find our own vocabularies affected by the forms of gigantism we set out to study. The fundamental methodological questions that accompany our investigation of gigantism's forms are therefore how to engage critically with the material and temporal categories that accompany gigantism and how to understand their interrelatedness when their effects underpin our very way of thinking.

Maria Finn, *Unfinished #18*, pencil on paper, 29 × 42 cm, 2018. © Maria Finn.

1   The Eiffel Tower: Grand-Scale Montage and Lightning Rod for Meaning

Prologue: An Elevator Ride

Paris, July 2016

It was the best of times, it was the worst of times. Incomparable but with an impact, it was the summer of pain: eighty-five people killed by a truck in Nice, an attempted army coup in Turkey, a man with an ax on a train near Würzburg, a dislocated right shoulder. With my left hand, I scrolled through Facebook updates from friends notifying me that they were safe in geographical locations that had become unsafe while I slept. The dislocation had made me strangely numb on one side, disturbing my sleep just as much as all the other things that were happening.

It was also the summer of latent feelings. I was having a love affair with a friend. Well, maybe not an affair in the strictest sense of the term because technically neither of us was cheating on anybody. Still, having known each other for years and being part of the same circle of friends, this was definitely a sea change, and we agreed it would be better to get it out of our systems, put it to rest. It. This thing. Whatever it was that had functioned latently as friendship, only to erupt at the most inconvenient of times, making a relationship I had deemed safe suddenly unsafe in a way so banal that Facebook had no button for it.

We left for Paris the morning after the Nice attacks and reserved a table at Le Jules Verne in the Eiffel Tower the same evening. The tower that Roland Barthes had called a lightning rod for meaning seemed an appropriate place to try to make sense of what had struck us, not like lightning—this was not love at first sight—but rather as the emergence of feelings that had remained unnoticed and materialized over time. Also, I was writing a book about towers with Henriette, and since neither of us had ever visited the Eiffel Tower, I had promised to go there on my own. It seemed the perfect hideaway to gain some perspective and to compress this emergent romantic potentiality into eighty hours in Paris.

Yet as I went up the tower that night, past the throngs of tourists waiting in line and into the lift that goes to the second deck where the restaurant is located, things were not coming together as I had hoped. As I looked out over Paris, through an iron construction that had seemed as transparent as a spider web from afar yet was brutally solid up close, the view disintegrated into a montage that pitted function against form. I was moving swiftly upward, the city disappearing below me, decomposed by the triangulation of iron girders that facilitated my ascent but obstructed the serenity of my view. The beams kept blocking a panorama of the city that would have allowed me a rational overview of the situation. Their interference revealed new angles of the familiar cityscape that were both titillating and uncomfortable. I took out my phone and started to film the ascent, asking my fellow traveler to step out of frame, yet feeling his breath on my forehead.

The restaurant was almost empty, apart from the elderly American couple seated next to us. We speculated that the attack in Nice had led to cancellations by the tourists who normally flocked to dine here, but we were unable to have this confirmed by the waiter. Later we were joined by a young couple celebrating their wedding anniversary and by a loud family whose toddlers transformed the restaurant into a playground. The passing of time seemed to merge into a field of simultaneity around us. After the meal, we went to the summit. It had darkened outside, and the tower was lit in the colors of the tricolor to honor the victims in Nice. We were caught in the fluorescent blue on the top third of the tower, looking out.

Back at our hotel, we could see the lights projected onto the tower. Seen from the hotel room window, the projection of the colors of the French national flag smoothed the tower's disjointed iron construction into a seamless image that made the tower seem as two-dimensional as the tricolor that our friends—whom we had tried to avoid by coming to Paris while we worked out what was going on between us—had layered on top of their Facebook profile pictures.

It is from this precise yet simultaneously highly mediatized point of view that this chapter approaches the Eiffel Tower, tracing some of its many different guises—as a privileged site for the romantic, a concrete object and tingling sensation, architecturally, affectively, and digitally, and as the multiplex, gigantic structure it has remained since its erection in 1889.

—KV

Going Up

> The tower attracts meaning, the way a lightning rod attracts thunderbolts; for all lovers of signification, it plays a glamorous part, that of a pure signifier; i.e., of a form in which men unceasingly put *meaning* (which they extract at will from their knowledge, their dreams, their history), without this meaning thereby ever being finite and fixed. Who can say what the Tower will be for humanity tomorrow?
>
> —Roland Barthes, "The Eiffel Tower" (1964)[1]

As the terror attacks in Paris on November 13, 2015, developed, they left a grim pattern of well-coordinated yet apparently arbitrarily connected urban sites, layering what seemed like a random topography of terror onto the city map. Whereas the World Trade Center attack of 9/11 had left the image of a frighteningly spectacular symbolic action that "came upon" the city as a violent downward movement, the Paris attacks reignited the latently distributed fear of terrorism as an attack that could erupt anywhere, at any time, in the midst of everyday urban life, and be felt by the citizens of Paris (along with those of many other cities in Europe and around the world). For two days following the attacks, the Eiffel Tower was closed. It reopened with a dazzling light show that used the tower itself as a projection screen for the French national flag. On social media, many people replaced their profile pictures with the French flag or an image of the Eiffel Tower, which in this way provided a vehicle for people to express sympathy for and even identification with France and the city of Paris. In its virtual installation, the tower appeared as a singular situated site that represented the events, as if the unfathomable acts themselves called for interlinkage with a symbolic site of extraordinariness.

Rather than signifying the technological advancement and striving for grandeur of late nineteenth-century French culture (Paris, the "city of light") or a tourist magnet for romantic getaways (Paris, the "city of love"), the Eiffel Tower now came to stand for and evoke a collective anxiety in the local context of France but also in the parts of the world where the magic of the Eiffel Tower as an emblem of the tourist experience of Paris was significant. The form of mourning that occurred in the aftermath of the Paris attacks and for which the Eiffel Tower became one of the main shared symbols is indeed marked by an intermingling of architectural and digital culture that is at the heart of what makes the Eiffel Tower culturally significant.

Moreover, the combination of architectural and digital culture that the digital rendition of the tower represents contains an implicit contradiction. The Eiffel Tower is curiously banal and unnoticeable in its ubiquity as a cultural cliché consumed by tourists in the French capital, but it also remains (and repeatedly resurfaces as) potent and wide-reaching, readily available for new meaning to attach itself to it.

In the prologue to this chapter, this contradiction is amplified. The tower is experienced as both a physical and a mediated site, the two seemingly blending into one. The prologue emphasizes how an experience of intimacy and affect can develop in the intersection between the architectural structure and its digital incarnations and that this intersection contributes meaning on its own account as well as providing a place from which to contemplate political dimensions of a particular situation.

In this chapter, we trace and historicize the way the intricate interweaving of architecture and digital culture manifests itself in the experience of the Eiffel Tower in relation to three different forms of gigantism—linear, semantic, and latent. This chapter therefore asks whether the Eiffel Tower might be not only a symbolic lightning rod that attracts meaning—a medium, as Roland Barthes argues, and an iconic montage structure that we zoom in on in the fragment in Maria Finn's drawing above—but also a literal, material lightning rod that embodies conflicting temporalities. Grown out of a Victorian culture where the Eiffel Tower may be said to have been a marker of progression tied to industrial culture, it epitomizes late nineteenth-century ideas of growth and progress. The Eiffel Tower was the highest tower in the world from its erection in 1889 until 1958. At that point, its vertical gigantism was superseded, although that did not seem to make it less iconic. It is as if the tower's gigantic impact—which started out primarily as a form of vertical, linear gigantism linked to its size—began to spread horizontally as the tower multiplied as a benign tourist sign, instating semantic gigantism. The most recent digital rendition of that semantic gigantism takes on more latent forms, whereby the tower's iconicity itself takes on new meanings that are more calmly present and simultaneously both more and less politically charged.

Lightning striking the Eiffel Tower, June 3, 1902, at 9:20 a.m. This is one of the earliest photographs of lightning in an urban setting. In Camille Flammarion, *Thunder and Lightning*, translated by Walter Mostyn (London: Chatto & Windus, 1906). © M. G. Loppé. National Weather Service (NWS) Collection, National Oceanic and Atmospheric Administration.

To trace these shifting forms of gigantism attached to the tower, we revisit the work of two canonical twentieth-century thinkers who have considered the Eiffel Tower and who are themselves no less towering than the building they write about: Walter Benjamin (1892–1940) and Roland Barthes

(1915–1980). In Benjamin's writings, the Eiffel Tower features not just as a tall structure hovering over Paris but also as an architectural montage with the potential to break open the linearity of history as progress. Barthes's reading of the Eiffel Tower, moreover, gives us a guided tour of how such a tourist site can evoke multivalent meanings for a mid-twentieth-century cultural theorist. Like Benjamin, Barthes points to ways in which vertical and horizontal forms of gigantism intermingle in the Eiffel Tower, in particular when he begins to question the clear dividing lines between the material and human in relation to the tower. The significance of these interdependencies mounts when discussed in light of the tower's installation as a collective site of mourning in and through the digital sphere in the aftermath of the terror attacks in Paris in November 2015. We therefore conclude the chapter with a discussion of the way the Eiffel Tower came to take on new latent meanings in a digitized rendition after that terrorist attack.

Montage as Method

From the outset, the Eiffel Tower was marked by a clash of expectations. When it was erected for the World's Fair in Paris in 1889 as a manifestation of the accomplishments of industrial culture and the progressive promise of the future, the Eiffel Tower was supposed to stand for only twenty years. Although it was Western culture's first fulfillment of the long-standing dream of building a 300-meter-high tower, as late as 1908 there were still plans for its demolition.[2] However, by then the tower had become important as a mast for telecommunications. In fact, shortly after its invention, a radio signal was transmitted from the tower's tip.[3] In 1909, an underground military radiotelegraphy station, which still exists today, was built close to the south pillar, and it was from the Eiffel Tower that a radio message was intercepted that gave the French army enough information about German troop movements to win the first battle of the Marne in 1914 and allegedly avoid an invasion of Paris. In 1957, a broadcasting aerial was added, bringing the tower to 324 meters in height and increasing its radio frequency range. This shift in use—from a physically high manifestation of the gigantic imbued with symbolic value to a structure whose continued existence depended on the invisible (yet no less gigantic) width of its radio frequency range—echoes a shift in emphasis from vertical to horizontal gigantism with respect to the cultural significance of the Eiffel Tower.

The Eiffel Tower was originally built as a symbol of late nineteenth-century technological prowess, attesting to modern building culture's ability to reach into the sky. From the beginning, however, this sense of peak was destined to be surpassed by future industrial advancements, and the tower was meant to be demolished. As the century progressed, this part of the tower's symbolic value became less and less significant. It moved from being a phallic entry in a height competition to being a container that carried visitors up and down from its viewing platforms, facilitating their gaze on the city. Every year, hundreds of thousands of visitors continue to take the elevator to the top and participate in turning the tower into the symbol of Paris that it is today.

Becoming naturalized as part of the historical city, the tower has taken on new meanings, but this elasticity of semiotic capacity has arguably served only to increase its cultural significance. Someone who picked up on this ephemeral quality of the tower at an early stage was the German Jewish writer Walter Benjamin. In the first half of the twentieth century, he was taking stock of landmarks such as the Paris arcades and other nineteenth-century ruins from his own notoriously eclectic perspective, which combined (among others) Marxism with Jewish mysticism. As a temporal category, the idea of progress places human capitalist production on a potentially never-ending upward curve, and this conception of history is what Benjamin sees being challenged from his vantage point in the early twentieth century.[4]

Perhaps surprisingly, since much of Benjamin's work is focused on Paris, the Eiffel Tower does not take up much space in his oeuvre and therefore in Benjamin scholarship. However, our aim here is to address the elusive role of the Eiffel Tower in Benjamin's writings as significant in its own right. We argue that the presence of the tower is woven into his texts, embedded in his tracing of the relics of modernity's advent in Paris, which Benjamin calls "the capital" of the nineteenth century. Indeed, as a gigantic structure that he sees as adhering to the logic of the montage, the tower parallels Benjamin's own method of assembling fragments of text, most notably in his gigantic *Arcades Project*, which amounted to over a thousand pages of quotations, annotations, and comments when it was first published posthumously in 1982.[5] Although the Eiffel Tower may appear too crude and obvious an example to take center stage in Benjamin's subtle mappings, it is undoubtedly there, looming over the city, emerging in the fragments of

The Arcades Project and in his musings on developments in engineering and iron construction.

One place where we do find the Eiffel Tower featuring centrally is in Benjamin's short text "The Ring of Saturn or Some Remarks on Iron Construction," which is believed to have been written in 1928 or 1929.[6] In this text, Benjamin traces the history of iron construction—from the luxurious design of winter gardens and arcades to covered markets, railway stations, exhibition halls, and bridges—and concludes the essay with the Eiffel Tower. The title indicates the double nature of iron, which he describes as simultaneously familiar and unfamiliar. Iron may seem as alien as the rings of Saturn, but at the same time we are as accustomed to it as we are to a balcony railing that has become so habitual that we are almost blind to its existence. Benjamin follows the process of coming to terms with the relatively new ubiquitous uses of iron as a building material, which in the nineteenth century was often camouflaged to resemble older and better-known materials such as woodwork. He identifies in that process a battle between the architect and the engineer. For Benjamin, "artistic" architecture is linked to the phantasmagoria of bourgeois capitalism, while "engineering" architecture is linked to social revolution.[7]

At the time of his writing in the 1920s, Benjamin sees the latest example of this cultural friction between art and engineering in Jugendstil or art nouveau, which he describes as "an effort to renew art on technology's own rich store of forms."[8] Here, the reaction is reversed: rather than turning to art and decoration to make ironwork look familiar, Jugendstil utilizes the possibilities available with the period's new materials and technologies. For example, wrought iron is used to envision new forms that draw inspiration from an organic vocabulary. The Eiffel Tower therefore not only embodies Victorian engineering work but also testifies to the principles of modern architectural construction before the introduction of architectural modernism in the early twentieth century—that is, before Bauhaus logics (which gained prominence in the architectural avant-garde in Europe in the period when Benjamin was writing) stripped buildings of their ornamentation and made function an aesthetic paradigm in its own right. Benjamin describes this tendency as a movement in architecture toward "porosity and transparency" and "the well-lit and airy" when contrasted with the historicist facades and closed interiorities of the Victorian bourgeois forms of housing with upholstered furniture and thick curtains in which Benjamin himself grew up and lived most of his life.⁹ The Eiffel Tower is lace and ornament. It is iron construction but also porosity, transparency, light, and air. It is firmly placed in Victorian culture and simultaneously points to new aesthetic and building paradigms. It is both a highly decorated whimsical structure and a completely transparent infrastructural artifact.

Benjamin articulates this ambivalence elsewhere in *The Arcades Project* when he quotes the well-known cultural-historical work *Kulturgeschichte der Neuzeit* (1927–1931) by the Austrian philosopher and historian Egon Friedell, who gives the following description of the Eiffel Tower: "It is characteristic of this most famous construction of the epoch that, for all its gigantic stature . . . it nevertheless feels like a knickknack, which . . . speaks for the fact that the second-rate artistic sensibility of the era could think, in general, only within the framework of genre and the technique of filigree."¹⁰ Although the tower was meant to stand in not only for Paris but also for France at the World's Fair, in line with Jugendstil the tower offers the expressiveness of a vitalist style that has no particular national or urban culture attached to it.¹¹ Instead, meaningfulness resides in the tower's triangulation of engineering feat and experimentation, laced with abstract ornamental stylistic references and a looming sense of anthropomorphism visible in representations from the period.

With its massive scale devoid of any particular explicit function, the task of building what was to become the Eiffel Tower—a task given to Gustave Eiffel, whose engineering company specialized in molded iron bridges—resulted in a structure whose main purpose lay in the pure expression of the cultural and technological prowess and progress of late nineteenth-century industrial culture.¹² The tower is built to contain and move people, offering an unprecedented use of elevators to take visitors up. It not only is conceived as an object that reaches for the sky but is a vehicle with the ambition of lifting humanity to new heights. In this sense, the architecture of the tower (unsurprisingly, perhaps, given Eiffel's background as a builder of bridges) may be seen as a material infrastructure of containment and as a medium of progression where the forward-moving flow of progress is epitomized by movement upward in a very literal sense.

The originating idea behind the tower is generally ascribed to the Swiss engineer Maurice Koechlin, who was chief of staff at Eiffel's company. Eiffel credited him with the idea, although there were later numerous arguments over whether Eiffel had in fact stolen the design from Koechlin.¹³ Eiffel

The Eiffel Tower

M. Eiffel, our artist's latest tour de force, June 29, 1889. © Sambourne, Linley, 1844–1910.

himself shot to fame with the building, and he became installed as a kind of Belle Époque starchitect surrounded by the scientific, political, and cultural glitterati of the time. He built a small apartment for himself atop the tower, a part of which is still preserved. Today it displays a wax figure of Eiffel sitting next to the American inventor Thomas Edison as if immersed in discussion, with Eiffel's daughter standing behind them. The apartment

Exposition internationale, 1889, Paris. © Bibliothèque nationale de France, 1889.

atop the Eiffel Tower also functioned as a science lab where significant scientific experiments—particularly those connected to radio and television broadcasting—would be conducted throughout the century. Eiffel met with world leaders and scientists in this space, which with its flowery wallpaper and painted wooden panels displayed the inward coziness of Victorian culture amid the exposed heavy steel beams of the tower.

In the ironwork text, Benjamin describes the time of the construction of the Eiffel Tower—some forty years previous to his own writing—as an instant when "that heroic age of technology found its monument in the incomparable Eiffel Tower."[14] By emphasizing its singularity, he makes the

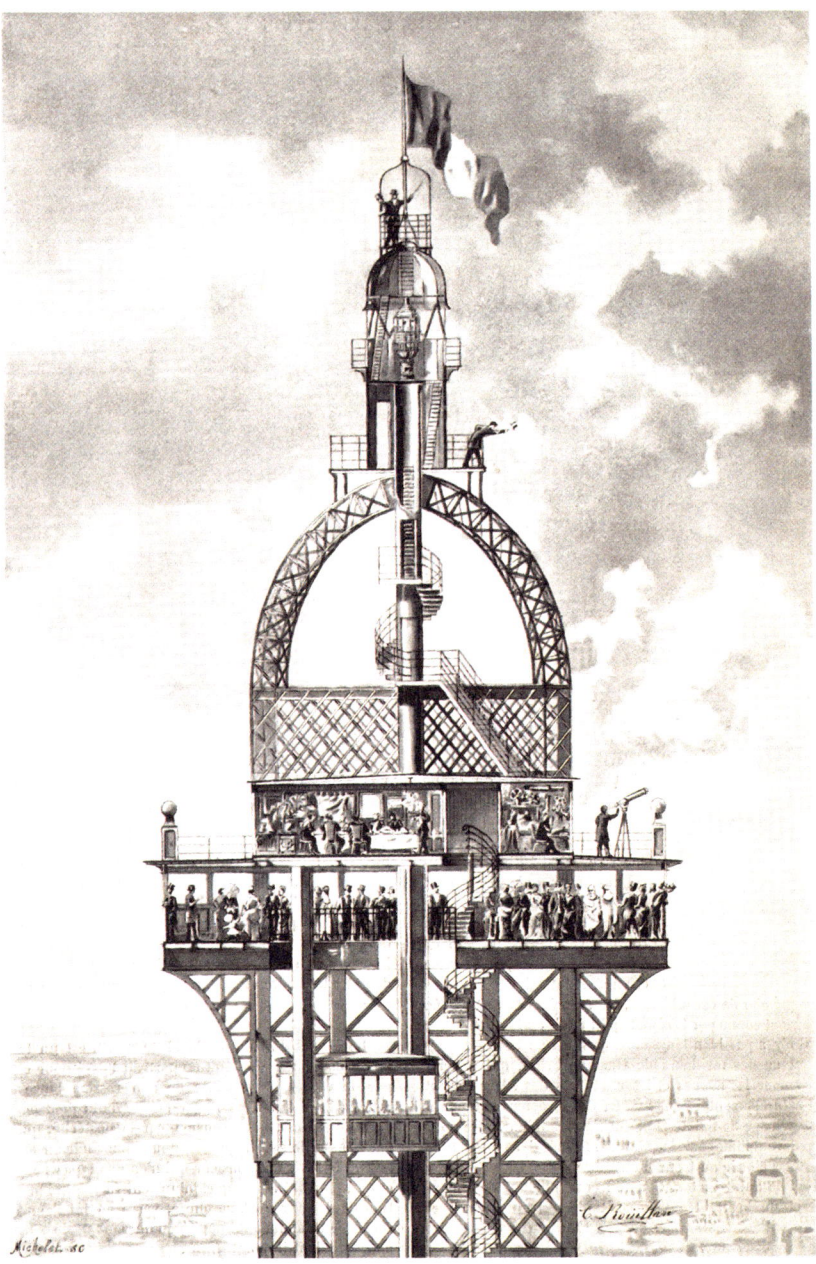

Exposition Universelle de Paris en 1889. La Tour Eiffel, le plus haut monument du globe. Imprimerie Typo-Lithographique. © Richet & Standachar (Clermont-Ferrand), 1889, Bibliothèque nationale de France.

tower more than just another example of iron construction. It takes on a significatory role as monument to an age. Unlike a monument to a battle or historical person, the Eiffel Tower bears witness to a shift in architectural aesthetics by physically embodying it.

In this way, Benjamin emphasizes a fundamental tension at the core of the tower's cultural significance: it testifies to an emerging tendency to expose the function of building materials and is a gigantic symbol in and of itself. The Eiffel Tower's "incomparableness" sets it apart from the other examples of iron construction that Benjamin mentions. However, the incomparableness also stems from the intended ephemerality of the tower as something that would be torn down and inevitably overtaken by something newer, taller, and even more gigantic. In this light, the tower's vertical gigantism comes to the test. It may seem that the logical thing would have been ultimately to demolish the tower and replace it with something better. But to this day, the Eiffel Tower has not been demolished. This fact inevitably changes the primary symbolism of linear progression that is attached to the design.

The narrative of the tower's resilience in the face of the threat of demolition gives the Eiffel Tower a very particular sense of cultural intent, as though the tower might have agency in and of itself. Herein lies a paradox. The tower *does* nothing: it simply stays standing. Rather than calling on an eruptive event, such as the tearing down of the tower and its replacement with a new and even more gigantic structure, the tourist industry throughout the twentieth century increasingly transformed the tower into a benign, harmless symbol. The explosive semiotic heritage—of progress, progression, and even revolution—of the tower was thereby held at bay. Ultimately, the Eiffel Tower was more and more a monument to itself, and it was repeated endlessly in other cultural media: in pictures, as figurines, in films, on postcards, in books. This process can be seen as part of a transformation of the tower's vertical gigantism into a more horizontal one.

In *The Arcades Project*, Benjamin preempts this process when he quotes historians Lucien Dubech and Pierre d'Espezel's writing on the Eiffel Tower from 1926: "Greeted at first by a storm of protest, it has remained quite ugly, though it proved useful for research on wireless telegraphy. . . . It has been said that this world exhibition marked the triumph of iron construction. It would be truer to say that it marked its bankruptcy."[15] The quotation expresses the idea that the Eiffel Tower marks the height of something (whether triumph or bankruptcy), but it also expresses the idea that the tower's original symbolism has given way to its function as an antenna used for the transmission of wireless signals. At the time Benjamin is writing, the Eiffel Tower has already shifted from being a symbol of the accomplishments of industrial culture (simultaneously loved and despised) to filling a practical function as a mast for communication technologies. Rather than becoming one of the urban ruins that intensely interested Benjamin (such as the arcades), the tower reinvents itself so that it is as much a vehicle for an emerging information society as it is a testimony to ideas about progress and progression. This transformation is central to the redemptive potential that Benjamin sees in the tower when he reads it as a manifestation of the principle of montage. As a technique for juxtaposing elements and sometimes found objects that would be pasted together to form a new image, the montage was popular among many avant-garde artists and writers in the early twentieth century, and the Eiffel Tower was represented in disjointed form in cubist painting. Architectural historian and philosopher Dalibor Vesely suggests that through the appropriation of the Eiffel Tower not just in the city but also "in paintings, films, literature, poetry, and music; and to some extent even in philosophy" (such as cubist paintings and Benjamin's writings), "the distance between its abstract appearance and the richness of the Parisian culture was articulated," and the gap began to close.[16]

The montage is indeed central to Benjamin's thinking, both as a topic of investigation and in the composition of his own writings. In "Envelope F" of *The Arcades Project*, which deals with iron construction, we see the way the montage is both a theme and formal characteristic of his own writing style:

> Never before was the criterion of the "minimal" so important. And that includes the minimal element of quantity: the "little," the "few." These are dimensions that were well established in technological and architectural constructions long before literature made bold to adapt them. *Fundamentally, it is a question of the earliest manifestation of the principle of montage.* On building the Eiffel Tower: "Thus, the plastic shaping power abdicates here in favour of a colossal span of spiritual energy, which channels the inorganic material energy into the smallest, most efficient forms and conjoins these forms in the most effective manner. . . . Each of the twelve thousand metal fittings, each of the two and a half million rivets, is machined to the millimeter. . . . On this work site, one hears no chisel-blow liberating form from stone; here thought reigns over muscle power, which it transmits via cranes and secure scaffolding."[17]

Champs de Mars: La Tour Rouge. 1911–1923, Robert Delaunay. Oil on canvas. © Robert Delaunay, The Art Institute of Chicago.

Here, Benjamin uses a quotation from German art historian Alfred Gotthold Meyer thereby bringing the Eiffel Tower into a discussion of montage as an aesthetic form. He thereby argues that this gigantic structure needs to be understood by scrutinizing its smallest parts: the vast number of quantifiable iron parts that make up the tower (approximately eighteen thousand). Its montage of building parts means that the tower is simultaneously both assembled and disassembled. This is what makes the tower resemble a montage as an artistic means of expression. Describing the Eiffel Tower as an architectural embodiment of the principle of montage, Benjamin links the montage to mechanical reproduction, a theme that he developed in relation to film. For Benjamin, techniques for mechanical reproduction that allowed the mass production of text, film, and other cultural artifacts also mark the fine line between catering for mass consumption (and the dangers of mass propaganda that come with that) *and* the potential to bring about a critical awakening.[18]

The montage is significant for Benjamin on a range of levels, from architecture to other aesthetic forms of representation. It is also closely linked to his own mode of writing, particularly the unfinished *Arcades Project*, the basis of which is to extract fragments of text from their original contexts and juxtapose them in new ways, creating constellations that would otherwise not be visible. Indeed, the massive extent of *The Arcades Project*—with its thousands of quotations, notes, and annotations from Benjamin's research—bears the hallmarks of gigantism, both linear and semantic. It is measurable and thereby linear but also points to semantic gigantism's preoccupation with the excess of meaning.

It is therefore key that one of the rare mentions of the tower in Benjamin's writings is when he notes that its montage technique forms a lens, which he likens to the task of the historian:

> "In the windswept stairways of the Eiffel Tower, or better still, in the steel supports of a Pont Transbordeur, one meets with the fundamental aesthetic experience of present-day architecture: through the thin net of iron that hangs suspended in the air, things stream—ships, ocean, houses, masts, landscape, harbor. They lose their distinctive shape, swirl into one another as we climb downward, merge simultaneously...." In the same way, the historian today has only to erect a slender but sturdy scaffolding—a philosophic structure—in order to draw the most vital aspects of the past into his net.[19]

Montage is a method for making visible the invisible processes that take place between discrete elements. It is a textual and conceptual form of scaffolding that carries a critical, progressive potential because it not only is spatial but also allows us to see what would otherwise be glossed over in the linear trajectory of time moving forward. The montage form is thus closely connected to Benjamin's objections to linear conceptions of history. Through the method of montage, whether textual or architectural, Benjamin conjures a representation of history and time as a total event that connects spatially rather than through linear untolding.[20] From the ideas of linear gigantism that governed the tower at its erection, Benjamin's thinking emphasizes the more horizontal ideas of spatial expansion and repetition that are also at work in the structure. The Eiffel Tower's constructional elements are a montage on a grand architectural scale. For Benjamin, the spatial principles of the montage technique point to a crisis articulation of central modern understandings of time as linear and as tied to hopes of growth and progression.

Benjamin can help us understand the difficult implications of what is meant with this shift from something spatial (a montage as a text, a building, or an image) to something that impacts our understanding of time. Benjamin regards history as neither a continuum nor a chain of events moving toward progress. Rather, he understands the past in flashes brought about by insights that can be generated through methods such as montage to create what he calls dialectical images: "It is not that what is past casts its light on what is present, or what is present its light on what is past; rather, an image is that wherein what has been comes together in a flash with the now to form a constellation. In other words: image is dialectics at a standstill. For while the relation of the present to the past is purely temporal, the relation of what-has-been to the now is dialectical: not temporal in nature but figural [bildlich]."[21] This implies a sense of momentary horizontal expansion in contrast to a more linear progressive history. For Benjamin, this indicates a temporal sensation of simultaneity whereby we find "what-has-been" embedded in the "now." Benjamin's eclectic theoretical sources of inspiration (including Marxism and Jewish mysticism) are here translated into a methodology that aims to tease out what remains dormant in the twentieth century from nineteenth-century bourgeois culture's dream imagery in which both architecture and technology have key roles to play.

Benjamin's allusions to the Eiffel Tower as montage emphasize a sliding movement from the time when the tower was primarily a celebrated symbol of the progress and progression of French and Western industrial culture to a time when it lingered as an industrial rupture in the historical urban landscape of Paris. The connotations of the montage endow the tower with the redemptive potential to expose the crisis of the dream imagery that it epitomizes. By regarding the tower—through Benjamin's writings—in this way, we can begin to see how the qualities that make it gigantic start moving from the vertical and physically imposing to a gigantism that is much more subtly and horizontally ingrained in the cityscape, in other media, and in the cultural imagination pertaining to early twentieth-century cities. If the tower is seen as a montage, its agency shifts from making its mark through its overwhelming presence to transforming its surroundings actively in a more indirect manner, and in this process it takes on an almost invisible quality. This shift indicates a semantic and spatial horizontal expansion where the tower comes to stand in for and even contain the surrounding city of Paris.

To further discuss this elusive quality of the Eiffel Tower, and the sliding movement from a vertical toward a more horizontal form of gigantism that is part and parcel of its cultural significance and the form of gigantism it comes to embody in the twentieth century, let us now turn to another canonical reader of the tower: Roland Barthes.

Seeing the Eiffel Tower See: Expanding Meaning

In his 1964 text on the Eiffel Tower, Barthes describes it as an architectonic structure often depicted in a mythopoetic language that organically connects it to the city of Paris, as though the tower could stand metonymically for the city. The perception of an overpowering and organic yet at the same time ambiguous relationship between city and tower was not new, and Barthes begins his text by referencing French author Guy de Maupassant (1850–1893), who often ate his lunch in the Eiffel Tower. For Guy de Maupassant, it was the only place in Paris where he was not forced to look at the tower constantly. Barthes continues:

Paris. Place du Trocadéro. 1961. © Lars (Lon) Olsson, 2011. Wikimedia Commons.

> It's true that you must take endless precautions, in Paris, not to see the Eiffel Tower; whatever the season, through mist and cloud, on overcast days or in sunshine, in rain—wherever you are, whatever the landscape of roofs, domes, or branches separating you from it, *the Tower is there*; incorporated into daily life until you can no longer grant it any specific attribute, determined merely to persist, like a rock or the river, it is as literal as a phenomenon of nature whose meaning can be questioned to infinity but whose existence is incontestable. There is virtually no Parisian glance it fails to *touch* at some time of day.[22]

The visibility of the tower is presented here as overpowering. The experience of the city is tied up with the physical presence of the tower, yet Barthes's text describes how the visual ubiquity of the tower also tends to make it invisible and causes it to blend into the surroundings. Because our gaze is constantly confronted with it, the tower becomes naturalized, like a rock or the river. It is not something that our perceptual apparatus needs to react to as a stimulus. The tower has passed into the habitual mode of our attention: we have become used to it to the degree that it has become unremarkable. However, being unremarkable in this context does not mean that it ceases to have an effect. Rather, Barthes picks up on the Eiffel Tower's capacity to take on new meaning and for meanings to multiply and expand.[23]

Barthes notes this capacity in the above-quoted text when he remarks that the tower's "meaning can be questioned to infinity." Whereas Benjamin described the Eiffel Tower as an "incomparable" monument, Barthes calls it a "useless monument" and an "empty monument,"[24] stripping it in this way of "deeper" potential symbolic meanings. Barthes narrates how Gustave Eiffel, confronted with the accusation that he had built a useless structure, defended it by pointing to its scientific potential for aerodynamic measurements, studies of the strength of materials, tests of the physiology of climbers, research in radio electronics, and telecommunication and meteorological observations. In this way, Eiffel emphasized the vertical gigantism as a symbol of scientific progress. Yet Barthes notes the futility of Eiffel's quest: "Eiffel saw his tower in the form of a serious object, rational, useful—men return it to him in the form of a great baroque dream which quite naturally touches on the borders of the irrational."[25] This irrationality marks a form of gigantism linked to scientific progress being slowly overtaken by another that is more elusive and moves horizontally in the form of a semantic widening.

One of the appeals that Barthes singles out and that seems to echo Benjamin's view of the Eiffel Tower as montage is the "surprise of seeing how this rectilinear form, which is consumed in every corner of Paris as a pure line, is composed of countless segments, interlinked, crossed, divergent."[26] Barthes, however, is not interested in cutting away the dream images that latch onto the Eiffel Tower. Instead, he traces the perspective and fascination of the tourist and compares them to a mystic ritual where meaning remains enigmatic. What makes the tower gigantic to him seems to lie precisely in this ephemerality. As a container, the tower does not offer the comfort of enclosed space for the visitor: it is both open and closed, function and ornament, and simultaneously marked by vertical as well as horizontal forces. Barthes writes: "One cannot be shut up within it since what defines the Tower is its longilineal form and its open structure. How can you be enclosed within emptiness, how can you visit a line? Yet incontestably the Tower is visited: we linger within it, before using it as an observatory. What is happening? What becomes of the great exploratory function of the *inside* when it is applied to this empty and depthless monument which might be said to consist entirely of an exterior substance?"[27]

For Barthes, the appeal of the empty sign lies not in its potential to reveal a deeper truth but rather in the freedom offered by the multiplicity of potential meanings that can be ascribed to it. Barthes is arguably a (post-)structuralist and not a modernist thinker. For him, plurality of meaning is key: there is always potential for new meaning to develop, nourished by the abstractness of the tower's exterior and the emptiness of its interior. What is key for Barthes is thus the semantic openness that this abstractness entails. He sees it as "a form in which men unceasingly put *meaning* (which they extract at will from their knowledge, their dreams, their history), without this meaning thereby ever being finite and fixed. Who can say what the Tower will be for humanity tomorrow?"[28] In this sense, the potentially endless play of signifiers that Barthes attaches to the Eiffel Tower comes to harbor a notion of futurity.

Each in his own way, Benjamin and Barthes bring to the table the cultural heritage of two of the twentieth century's significant European theoretical movements—the Frankfurt School and French poststructuralism. The juxtaposition of these canonical interpreters of the Eiffel Tower reveals the shifting connotations of the way in which the architecture of the tower is intertwined with its cultural significance and meaning. Yet the shifts in emphasis of these two readings say as much about the Eiffel Tower itself as about the shifting tenets of twentieth-century cultural theory.

Embrasement de la Tour Eiffel pendant l'Exposition universelle de 1889. Georges Garen, 1889. © RMN-Grand Palais (Musée d'Orsay) / René-Gabriel Ojéda.

It is therefore significant that in the same way that Benjamin's reading of the Eiffel Tower's montage-like qualities resembles the montage technique he employed in his own writings, Barthes's argument about the Eiffel Tower's hybridity is itself highly multivalent. Notably, Barthes pinpoints the embeddedness of the tower in the cityscape and the way its naturalization is part of what makes it an open signifier, attracting meaning. For him, the architectural structure is endowed with the ability to reach out and touch—not the human body or even the eye but the "glance" (*regard*). He generates an understanding of the tower as something that tries to contact its surroundings, as can be seen in images anthropomorphizing the tower's presumed gaze. Barthes's reading of the tower as an architectural structure frames it as a transmitter or communication channel that has different modes for reaching its surroundings. He writes:

> The Tower is an object which sees, a glance which is seen; it is a complete verb, both active and passive, in which no function, no *voice* (as we say in grammar, with a piquant ambiguity) is defective. This dialectic is not in the least banal, it makes the Tower a singular monument; for the world ordinarily produces either purely functional organisms (camera or eye) intended to see things but which then afford nothing to sight, what *sees* being mythically linked to what remains *hidden* (this is the theme of the voyeur), or else spectacles which themselves are blind and are left in the pure passivity of the visible. The Tower (and this is one of its mythic powers) transgresses this separation, this habitual divorce of *seeing* and *being seen*, it achieves a sovereign circulation between the two functions; it is a complete object which has, if one may say so, both sexes of sight.[29]

In his efforts to describe the multiple meanings of the Eiffel Tower, Barthes pitches the tower as an architectural structure that can be read simultaneously as agent and object. He detaches the Eiffel Tower from a dialectic between the hidden male voyeur (but also a phallic tool) and the exposed passive female object (and thus a container). Instead, Barthes argues for the singularity of the Eiffel Tower as a hybrid object insofar as it is simultaneously observer and observed, male and female, agent and object. Barthes thereby collapses the dualisms that attach to the tower's vertical gigantism into a more hybrid form; and in this collapse, a new semantic openness is created that gives rise to new meanings.

In the time since Barthes wrote his text in 1964, the dissolution of dichotomies such as those mentioned here has been a huge concern across many cultural-theoretical movements, and as the gendered metaphors that

Barthes calls on point to, feminist theory and queer theory have been pivotal. Barthes here establishes the Eiffel Tower as a material structure whose capacity as a medium is to transgress the usual dichotomies between male and female, human and machine, medium and meaning. This line of thought resonates with Donna Haraway's famous "A Cyborg Manifesto: Science, Technology, and Socialist-Feminism in the Late Twentieth Century," a text that can be read as vesting a utopian hope in the hybridization of humans and technology but that also contains a subtle decentering of the human as the sole agent of action and meaning. It explores differentiated relationships between humans, technologies, and material culture in a way that is specifically targeted at repudiating typical modern dualistic concepts:

> To recapitulate, certain dualisms have been persistent in Western traditions; they have all been systematic to the logics and practices of domination of women, people of colour, nature, workers, animals—in short domination of all constituted as others, whose task is to mirror the self. Chief among these dualisms are self/other, mind/body, culture/nature, male/female, civilized/primitive, reality/appearance, whole/part, agent/resource, maker/made, active/passive, right/wrong, truth/illusion, total/partial, God/man. . . . High-tech culture challenges these dualisms in intriguing ways. It is not clear who makes and who is made in the relation between human and machine. It is not clear what is mind and what body in machines that resolve into coding practices. In so far as we know ourselves in both formal discourse (for example, biology) and in daily practice (for example, the homework economy in the integrated circuit), we find ourselves to be cyborgs, hybrids, mosaics, chimeras.[30]

Indeed, the anthropomorphic Eiffel Tower can be seen as a cyborg, and such a reading situates it between the human, the technological, and the architectural. Taking into consideration Haraway's ideas of a feminist posthumanism of solidarity between species, it seems obvious to note that it is not only men who "unceasingly put meaning" in the form of the tower, as Barthes writes. Similarly, where Barthes asks "what the Tower will be for humanity tomorrow," Haraway would most likely encourage us to engage a wider spectrum than the implicit emphasis on human culture that Barthes engages here.

However, in exactly that move of relational posthumanism, another more latent gigantism—to which we shall now turn—might in fact rear its head. To engage this latent gigantism, we may ask what happens if, rather than seeing the tower as a rectilinear infrastructural whim reaching for the sky, we instead see it as a leaking container? What happens if, rather than seeing the tower's overpowering iconicity as a semantic form of gigantism, we instead regard its malleable sign quality as a form of gigantism that operates more latently? To do so, we now turn to a more recent chapter in the tower's history—its digital life as an image in the aftermath of the terror attacks on Paris in 2015.

Leaks, Crisis, and Gigantism

First, let us think back a couple of years to the days immediately after the terrorist attacks in Paris in November 2015. For a Copenhagen-based social media user from a Scandinavian, academic demographic, the newsfeeds during the days after the Paris attacks seemed to overflow with declarations of sympathy for the victims of the attack, expressed in (among other ways) the layering of profile pictures on social media inside and outside France with the French national flag and images of the Eiffel Tower. As is discussed at the beginning of this chapter, this viral adaptation of the Eiffel Tower was an act of solidarity. An example is a drawing fusing the tower with the peace symbol, which was posted on Twitter by French graphic designer Jean Jullien in the hours after the attack.[31] His drawing compresses the approximately eighteen thousand metal fittings and the two and a half million rivets that make up the Eiffel Tower into a single smudged black line, which makes the tower look like a peace sign. In its simplicity and 1968 nostalgia, this straightforward visual strategy and its message are apt for viral dissemination. Yet the context of violent attacks and expressions of solidarity indicates that the connotations of the Eiffel Tower we saw at work in Benjamin's and Barthes's readings have undergone further transformations and taken on new meanings. The present appropriation of the tower on social media, where the Eiffel Tower still exudes a gigantism, is one, as we argue, in which the iconicity of the architectural form of the tower has become co-opted in this context for strategies of meaning making that—although no less gigantic—operate in ways that are more latent. These latent strategies are characterized by a particular form of flattening that is tied to the abstraction of the interfaces of different forms of social media and digital communication platforms, which helps make the tower available to appropriation for widely diverging purposes.

A few months after the Paris attacks, in the spring of 2016, the Eiffel Tower was lit up again, this time in the colors of the Belgian flag. Once

The Eiffel Tower's transmitting antenna. Top floor of the Eiffel Tower, Paris, Region of Île-de-France, France. April 7, 2017. © Zairon, 2017. Wikimedia Commons.

Jean Jullien, *Peace for Paris*, paint on paper, 2015. © Jean Jullien, 2015.

again, the Eiffel Tower manifested itself as a vehicle for the expression of sympathy and grief, this time with a neighboring country, and the image of the tower clad in the Belgian flag went viral on social media. By assuming the colors of another nation-state, the tower transgressed its status as a national emblem. It became a symbol of Paris and France reaching out horizontally toward its European neighbor, Belgium. In this way, the tower's open structure assumed screenlike qualities. It emanated light messages unrelated to its original physical appearance as a symbol of the accomplishment of industry, and it did so in a way that went beyond the abstraction of the open signifier and reverted to an easily decodable symbolic gesture, albeit with no less grandiosity in its symbolic language or impact. The Eiffel Tower's visual form seemed to be perfectly crafted for global viral dissemination while simultaneously also opening itself for a critique of foregrounding the grievability of European bodies. The tower thereby became a synecdoche not of tourism but of crisis articulation as a universalizing grand gesture that makes apparent the anachronism of such a gesture in a global context marked by inequality of grievability.

This connotation is taken to its extreme in a warning video disseminated online in summer 2016, generally attributed to members of the militant group Islamic State of Iraq and Syria (ISIS). The video shows the Eiffel Tower falling to the ground in a way reminiscent of what happened to the World Trade Center on 9/11.[32] This imagery along with Jullien's drawing can be seen as examples of how the Eiffel Tower and its charged history can be co-opted for widely disparate agendas that engage with and rearrange its symbolic heritage. Indeed, a significant connotation of the Eiffel Tower is that of revolution. Erected in 1889, a year that marked the centennial of the French Revolution, from day one the tower was a symbol of change, its height denoting that something had been overthrown in order for the structure to shoot up.[33]

Despite their fundamental differences, both Jullien's sentimental peace symbol drawing and the digitally edited video portraying the tower as a potential target are constructed with global dissemination on social media platforms in mind. What they have in common is that they unsettle the iconicity of the tower's rectilinear architectural form, either by turning it into a peace emblem or by suggesting it should be destroyed. Both ways of reconfiguring the tower's vertical gigantism into an easily digestible visual message (calling for peace or for destruction) are indebted to the affordances of the social media platforms on which the messages are distributed. As media theorist Eric Snodgrass has argued with regard to Facebook's information architecture for content generation, the interface foregrounds a certain abstraction and "imposes its aesthetically homogenous, unalterable and clinically clean frames in a way that flattens, or at least drastically mutes, any visual sense of subversive intention and expression on the part of the creators of such pages."[34] Such abstraction is what enables the iconic form of the tower to spread horizontally—not as a semantic widening, in the sense of a free play of the tower as signifier, but in a condensation of the tower's potency—into an alluringly simple message that asks to be shared, ideally to go viral and amplify its reach, and that places less emphasis on whether the content summons collective mourning or incites revolt.

This can be regarded as a form of gigantism where the accumulation and calculability of reach and likes become qualitative measures in their own right and where there is a need to rethink categories such as the global and the collective in this context. The digital platforms can be said to reinforce a sense of mediated commonality—for example, by facilitating the massive number of times the different image renditions of the Eiffel Tower are shared. But at the same time, they potentially contribute to silo formations of audiences or to the risk that those who "share" and "like" these renditions become subject to data profiling and predictive analytics. Networked digital media enable a particular form of mediation between possibly globally linked but vaguely articulated potential collectives of "people like you" and the individual participating in informational flows. This situation often renders that individual an active (although often unknowing) part of a larger common order of resemblance—which can be understood as a categorization not based on traditional markers such as class, gender, or race but based on how groups of individuals act online.[35] Its gigantism is thus of a different and much more latent kind.

To understand what is latent about this situation, let us consider in more detail how the way in which information spreads implicates its users, either by providing an impetus to respond through liking, sharing, and commenting or through the data mining that such acts facilitate. In the context of a study of bullying and sexual harassment of young women online, Wendy Hui Kyong Chun and Sarah Friedland describe these characteristics as a process of leaking:

> New media are not simply about leaks: they are leak. New media work by breaching, and thus paradoxically sustaining, the boundary between private and public: from the Internet's technical protocols to its emergence as a privately owned mass medium, from social media's privatization of surveillance to its redefinition of "friends," new media compromise the boundary between revolutionary and

conventional, public and private, work and leisure, fascinating and boring, hype and reality, amateur and professional, democracy and pornography.³⁶

This leaking quality is at the heart of the transmission of networked media: leaking is not accidental, nor is it the sign of a fault. A network—of people, institutions, or machines—*has* to share information among its members, ideally at high speed, in order to sustain itself. Networks *have* to be connected, and they *have* to transmit the information they contain. However, a culture that values safety, security, and secrecy conceives of these technologies as "deceptively protective," Chun and Friedland argue, which makes the act of transmission riddled with danger and construes the leak as a potential disruptive threat despite its equally valid connecting function.³⁷

In contrast to the classic forms of spectatorship in post–World War II media such as television, networked digital media entail the possibility of personal involvement and agency. On social media we can interact, but even if our interaction amounts to merely scrolling the newsfeed to bring up new information, we may still *feel* that we take part. The possibility of individual responses produces the potentiality at least to take part in a particular situation by reading about it or more actively by reposting news or engaging in discussions online that may endlessly multiply and that produce movement and flows of data that are constantly tracked and monetized for data-mining purposes.³⁸ Despite these latent measures of surveillance and profit generation, the opportunity to respond and participate contributes to the impression of a distribution of the authority of who gets to disseminate news and decide what constitutes a crisis. Yet this process also contributes to generating a perpetual sense of crisis that undermines the impact of any gigantic event or catastrophe in its singularity. Each individual crisis can still be understood by calling on the rhetoric of the singular gigantic catastrophe in the way it implores us to act by clicking, sharing, or reposting on these digital platforms of networked media. This point also affirms a horizontal gigantism of a continual sense of constantly unfolding crises in the plural, which operates despite the fact that networks always function unequally, positively serving or harming some people more than others.³⁹

Whereas the Eiffel Tower originally connoted revolution as the exceptional overturning of the existing order, the tower in this rendition seems to point toward a sensation of perpetual crisis to which people all over the world (within the paradigm of a global connected village) can relate and respond instantaneously through social media sites such as Facebook, Twitter, or YouTube. Chun makes an astute argument regarding how the notion of crisis has changed in a new-media context which is helpful for considering the sense of latent gigantism that comes with it:

> Crises have been central to making the internet a mass medium to end mass media: a personalized mass device. . . . crises—moments that demand real time response—make new media valuable and empowering by tying certain information to a decision, personal or political (in this sense, new media also personalize crises). Crises mark the difference between "using" and other modes of media spectatorship/viewing, in particular "watching" television, which has been theorized in terms of liveness and catastrophe. Comprehending the difference between new media crises and televisual catastrophes is central to understanding the promise and threat of new media.[40]

If we think about the word *crisis*, the notion of crisis in and of itself makes sense only within a particular understanding of time, which is linear in orientation.[41] A sense of crisis emerges when a situation does not meet one's expectations of the future—for example, if we expect forward movement and progress, and this expectation is not met. A notable recent example is the so-called financial crisis of 2007 and 2008, which was experienced as a crisis because the economic recession to which it led circumvented the assumption (within the widespread horizon of expectation that characterizes contemporary Western culture) that future economic growth was a desired given. German historian Reinhart Koselleck links this understanding of crisis to a particular historical point in time in the nineteenth century when understandings of time and history as linear and as progressing into a future dominated. This understanding sets up a horizon of expectation that points into the future, and it thus allows an understanding of crisis to emerge.[42] Seeing humans as agents that are able to influence culture and change the future for the better is not just a question of projecting expectations into the future. It also is linked to an understanding of progression and progressiveness. A sense of crisis emerges at the very moment when such expectations are refuted and rebuked.[43]

In contrast, the leak is a conceptual figure that calls attention to the dynamics behind the idea of a permanent state of disruptive crisis and the temporality that goes with it, which does not conform to notions of linearity. The sense of a state of constantly unfolding crises makes the revolutionary habitual, not least because it points to the flattening of what constitutes an event that elicits a response. In the context of networked media, there almost seems to be a moral imperative connected to responding that in part arises out of the way digitally designed platforms can incite addiction and that is particularly effective on guilt-ridden consumers who feel a responsibility to counter mounting crises such as climate change through everyday actions.[44]

The classic notion of crisis, which is tied to the idea of the linear forward movement of time, is thus challenged when the underlying temporal understandings of progress and linearity are contested by other temporalities—for example, the idea of a broadening present. This comes to mean that we are dealing with a more habitual notion of crisis that is constantly and yet latently present and that we accept, often driven by what literary and cultural scholar Lauren Berlant in her theory of affect calls "cruel optimism"—which concerns, for example, fantasies of the good life that keep people living within a situation of crisis.[45] It is this sense of constant lingering crisis as a condition that networked digital media simultaneously produce, respond to, and feed into—for example, in measures to predict and preempt crisis through networked analytics that fold an anticipation of risk

La tour Eiffel illuminée en bleu blanc rouge—Fluctuat nec Mergitur—Liberté, égalité, fraternité, November 21, 2015. © Yann Caradec, 2015. Flickr. (CC BY-SA 2.0).

and uncertainty back onto the present and that emphasize a latent form of gigantism.

In both Jullien's and ISIS's digital renditions of the Eiffel Tower, we find a (re)politicization of a piece of architecture that has been replicated as a sign throughout its history, to the point where the original intention behind the building seems almost fully exhausted. Yet in the transposition of the tower to the digital realm, the architecture is invested with new symbolic power. This symbolic power does not only hinge on the tower's physical presence and the symbolic significance invested in its iconic vertical form. It also depends on more intangible and invisible infrastructures that are personally and globally interlinked and distributed at the same time as they are imbued with participatory potential.

In such digital reincarnations of the tower (and in the way it engages with a sense of immediate crisis), we see evidence of an intermingling of vertical and horizontal gigantism that is both gigantic and latent—gigantic in its presence and latent in its reach. As a gigantic structure, the Eiffel Tower has become an apt vehicle for a form of digital appropriation that feeds off its properties as a lightning rod for meaning. Yet this digital appropriation reaches beyond poststructuralist significatory play and points to a more fundamental ontological flattening—as we have seen here, for instance, in the question of what constitutes a crisis and who gets to decide when the nonhuman agents of algorithms also are part of determining the reach and range of the imagery. Since it was built and up until today, the Eiffel Tower has remained a significant—although not semiotically stable—cultural symbol that weaves together architectural, technological, cultural, symbolic, and political ideas. From the fusion of these ideas in its gigantic montage-like form emerges the appropriation we see in digital, networked media. Benjamin picked the tower's modern constructional principles apart into its smallest bits. Barthes saw it as a whole emptied of meaning. In its early twenty-first-century digital incarnation, the Eiffel Tower becomes an imprint of latent gigantism.

The concept of the leak construes networked media as giant containers built to leak. When the Eiffel Tower in its viral incarnation becomes one of these leaked pieces of information, its particular gigantism—which we have traced in this chapter from the linear to the semantic—takes on latent qualities. The global reach of the Eiffel Tower hinges not on how far afield the built structure is visible in the physical landscape but on the range of the tower's digital distribution, which, in a curiously anachronistic way, comes to foreground a perception of Europe as the center of the world. The Eiffel Tower can in this way itself be seen as a leaking container—one that, by virtue of its montage form and its open transparent structure, contains several forms of gigantism that seep from it, thereby enabling what we call its latent gigantism.

Maria Finn, *Unfinished #20*, pencil on paper, 29 × 42 cm, 2018. © Maria Finn.

2 The Twin Towers: The Remanence of the Twins

Prologue: A Telephone Conversation on 9/11

Copenhagen, September 2019

It was afternoon in Copenhagen on September 11, 2001. I called Kristin from inside a shopping mall—probably to convey a practical message concerning the study program we both attended at the time. I must have been using my mom's cell phone as I didn't have my own cell phone at the time. Kristin told me that planes had crashed into a building in New York City but as I couldn't quite picture what she meant, I continued shopping after we said goodbye on the phone. When I later saw footage of the terrorist attack on the news and realized the severity of what had happened, I watched the news in disbelief but also with a nagging feeling that I had not reacted appropriately to Kristin's message. The events of that somber day would have a massive impact on global politics—and on many people's lives.

The media backdrop I recall most vividly from this period is a very real but dismembered body—not the body of one of the victims of the attack but the characteristic and strangely apathetic movement of George W. Bush's mouth whenever I saw him speak on television. I recall a strange apparition of George W.'s staring eyes and moving mouth, but I have to concentrate hard to remember more complex facial or bodily features. I also recall the stale look on Bush's face as he sat in a classroom listening to schoolchildren learning to read, when the White House chief of staff whispered the news of the second plane crashing into the World Trade Center into the president's ear. The famous eleven words spoken on the eleventh of September: "A second plane hit the second tower—America is under attack." The words were followed by seven minutes of calm—seven minutes when the president stayed in the classroom and listened to children reading, later stating, in a controversy-provoking statement, that, in the moment's bewilderment, his priorities had been not to upset the children and to project strength and calm to the crowd in the room and the cameras. Another image that pops into my mind is a grainy camcorder film of one of the towers, with a tail of smoke emanating from around the ninetieth floor. This image looks like a cinemagraph—a still picture with only one or two elements that move—this time of a city of latent turmoil where the only thing that moves is smoke.

September 11, 2001, was not the first time the Twin Towers had left a perplexingly disconnected imprint on my imagination. As a teenager on a visit to New York City, I recall passing the World Trade Center on a "hop-on, hop-off" tourist bus and thinking: "Hey, what's the fuss about these buildings?" From a ground-level perspective, they stood out as flat, unwelcoming facades. I never took an elevator up to the viewing platform, although I could easily have done so. In the week following the 9/11 attack, my architecture history teacher at university, with her usual sense of drama, dressed in black, and her pale, tragic features became an inverse mirror image of my own distance from the direct experience of bereavement. Another teacher talked in an overzealous tone about the attacks as the culmination of the "space race" and as evidence that this upward-reaching race would enter a new phase through terrorism in the twenty-first century. Yet the 9/11 attack seemed to me to be just as much about horizontal as vertical structures, and back then, reading Rem Koolhaas's *Delirious New York: A Retrospective Manifesto for Manhattan* for the first time made me think that the Manhattan grid was more robust than any building it accommodated.

Today, in every film or image of the Manhattan skyline I see, I still quickly scan the picture for evidence of the towers. The precise moment of the buildings' collapse reverberates in my memories, and when we return to this site in this book, we engage with the diffuse atmosphere surrounding my brief telephone conversation with Kristin on September 11, 2001. It was a context characterized by conflicting and contradictory sentiments. It soon felt clear that the minutes of the attack marked a juncture, although we did not know how to understand what that juncture meant. Although an increasing number of people today did not experience or were too young to have any recollection of the event, the towers have a lingering presence as a silently twittering other in many people's understanding of New York City. This latent presence has little to do with narratives claiming degrees of authenticity of the experience of 9/11, nor does it concern the direct implications for urban and world politics in the aftermath of the attack. Rather, it speaks to a continual engagement with the phenomenon of the sudden disappearance of two gigantic towers.

The Twin Towers

towertotower I sit at the back of the picture, slumping, wearing 😎 The Twin Towers seem to reach into the sky #memory #phonecall #twintowers #scalesofimpact

Engaging with their haunting and haunted aspects nineteen years later, this chapter works through the different forms of gigantism that the Twin Towers embodied and engages with the remanence—the residual magnetism—of the buildings that remains today.

—HS

Scales of Impact

A report from 2003 carried out by the Committee on the Internet under Crisis Conditions: Learning from September 11, a project sponsored by the National Research Council, sums up the events of September 11, 2001, with a focus not on the vertical collapse of the towers or the people who were tragically contained in that downward movement but on the horizontal impact of the destruction of the Twin Towers on the telecommunication network. Given that the towers are nodal points in a larger communication infrastructure—both through the visible antenna on the roof of the North Tower and through more invisible infrastructural wirings in and around the Twin Towers themselves—the collapse had effects that spread as far as South Africa and Romania. From this perspective, the events of 9/11 are described as follows:

8:45–10:00 a.m. Towers are attacked and set afire. Interior World Trade Center (WTC) communication is disrupted. Increased volume congests local exchanges and wireless networks. Limited physical damage occurs to the surrounding local telephone networks.

10:00–11:00 a.m. Towers collapse. Because the WTC was a significant wireless repeater site, some wireless connectivity is disrupted (Sprint PCS, Verizon, AT&T Wireless). Several ISPs' points of presence (POPs) in the complex—those of WorldCom, AT&T Local Service, and Verizon/Genuity—are destroyed. Some data and private-line services to a diverse set of customers in New York City, Connecticut, Massachusetts, and even some European locations are disrupted.

11:00 a.m.–5:00 p.m. Local power failures occur and some equipment is switched over to battery and/or generators. Fires burn in the WTC complex.

5:20–5:40 p.m. WTC Building 7 collapses, destroying a Consolidated Edison electrical substation in the process. The collapse also breaches the 140 West Street Verizon central office building, causing damage to equipment and the flooding of basement power systems. The fires, collapse, and flooding knock out much of the telecommunications service in Lower Manhattan.[1]

The Twin Towers

The report portrays New York City as a superconnected node due to the number of internet users, private data networks, internet service providers, and fiber-optic grids that come together in Manhattan. Here, internet providers connect with each other in so-called carrier hotels—buildings where telecommunication carriers lease space in order to link with other carriers in the same building. In this sense, New York City can be seen as a central intersection of vertical and horizontal information flows that were disrupted by the attack.

The central offices of the dominant local carrier on lower Manhattan, Verizon at 140 West Street, were severely impacted when across the street, Building 7 of the World Trade Center collapsed, crushing walls and cable vaults and flooding offices. This caused some fourteen thousand business and twenty thousand residential customers to lose their phone service, and it severely disrupted data communications.[2] Although the fiber-optic infrastructure could recover from the collapse of the Twin Towers by routing around the damaged systems, the infrastructure did not have capacity to "self-heal" in light of the damage that occurred when the Verizon building was also impacted. Also, the event itself caused internet traffic to increase

The Twin Towers from Greenwich Street after United Airlines flight 175 crashed into Two World Trade Center. © Hans Joachim Dudeck, 2009. Wikimedia Commons.

to a level where its capacity in 2001 was severely overburdened. Google reportedly issued a statement online that said: "If you are looking for news, you will find the most current information on TV or radio. Many online news services are not available, because of extremely high demand."[3] Moreover, indirect effects were caused by power failures. Although backup batteries and generators were activated, several providers did not expect longer power outages, which led to perturbations in internet connectivity when the backup batteries ran out of power on the morning of September 12. Further, there were longer-term effects caused by the overheating of network equipment located close to the World Trade Center site. The report from 2003 anticipated that this equipment would be less reliable in the future, although it was still running after the event.[4]

The concrete damage to the skyscrapers thus had horizontal infrastructural reverberations that reached far beyond Manhattan, the impact of which seemed to have a certain inbuilt latency that would make some disruptive effects felt later. As the title of the report *The Internet under Crisis Conditions: Learning from September 11* indicates, the word *crisis* in this context refers to a state with a prolonged temporality that stretches beyond an event related to a specific date. There might be latent effects that show up later, and the mission is to extract learning points that can be used to reduce the impact in the event of future emergencies.

On a global scale, the collapse of the Twin Towers caused visible effects on global routing. These effects sometimes occurred in unexpected places: the structure of connectivity does not adhere primarily to distance but rather to treaties, historical ties between countries, and geography. For example, it is easier to run a cable under water than across land. New York City is therefore a key nodal point for telecommunication infrastructures, not just for North America but also for Africa and parts of Europe. This explains why networks were affected on these continents when the towers were destroyed. At the same time, the report mentions that overall, as a gigantic system, the internet was more stable than usual on September 11. The report speculates that a reason for this might be that network operators generally tend to avoid optional maintenance as well as hardware and software changes during emergencies. Also, anecdotal information suggests that network operators were watching the news rather than making scheduled changes to their routers on 9/11.[5] Such reports are reminders that as an infrastructure, the internet is not neutral: local, structural, and human factors impact its functionality in significant ways.

This curious situation of global network stability during and after the attacks—starkly contrasting with the damaged buildings and infrastructure and the monumental political and symbolic impacts—shows that a crisis can manifest itself in different ways, with different degrees of latency, and on different scales. If we zoom in locally at the World Trade Center site of the attack on Manhattan, the impact is on a monumental scale of total destruction, and it carries grand symbolic connotations of the catastrophic that stem from the destroyed buildings' gigantism. We see a link to the symbolism tied to ideas of freedom and progress inherent in the idea of the American dream symbolized by the Statue of Liberty, as captured in Maria Finn's drawing at the start of this chapter. But when seen on a global scale in relation to telecommunication infrastructure, where the World Trade Center's antenna, servers, and cables combined to make the towers nothing but a node in a vast infrastructure system, the event lingers latently and can essentially go unnoticed. This paradox prompts a reconsideration of the relationship between local and global impact factors in which the local is often intuitively associated with smaller-scale effects. It shows that the question of what makes something gigantic in scale is as complex and historically contingent as the question of what makes it possible to name something a crisis, a question we broached in the previous chapter.

Moreover, feminist scholars remind us that the rhetoric of scale is heavily politicized and often gendered.[6] As an event, 9/11 can be said to have been planned, executed, responded to, and later analyzed in a vocabulary of the gigantic aligned with militarism and hard masculinity. As several commentators have pointed out, men—hijackers, rescuers, national security officers, media commentators, and scholars—filled the news, while women seemingly disappeared from view. Another example is that only two out of fifty opinion pieces in the *New York Times* newspaper in the first six weeks after the attack were written by women, and when women appeared in the news coverage, it was primarily as victims or relatives of victims.[7]

This situation feeds into feminist critiques of how women's experiences often are historically, socially, and culturally connected with an intimate, local, and bounded private world, while masculinity is more often connected to an expansive, political, and aggressive field of action. But it also reflects the well-known feminist argument that even the most private realm is conditioned by politics. As geographer Geraldine Pratt and literary theorist Victoria Rosner argued in 2002, the local-global binary and its gendered parallels need to be challenged insofar as intimate relations have both local and global effects and conversely global phenomena also make their mark in the intimate realm.[8] This is a claim to which the prologues to both this and the previous chapter speak. As Jan Jindy Pettman has argued, feminist perspectives on September 11 are able to "complicate, internationalize and gender the account."[9] We may use this as a guiding parameter to question the equation of vertical gigantism with phallic and even militant masculine ambitions of global impact. It also can be used to nuance our understanding of the gigantism at work in the scope of the attack on September 11, in the buildings that were struck that day, and in the cultural-theoretical responses they have solicited.

In this chapter, we contribute to an analysis of the politics of scale in the context of 9/11 by unraveling the different forms of gigantism at work in the Twin Towers. We read these buildings not only as phallic symbols of accomplishment and dominance but also as gigantic mediating structures and containers of people, wealth, information, inequalities, death, and bereavement. The Twin Towers were both vertical and networked giants, iconic architectural forms as well as hubs for information infrastructures. Moreover, as we have indicated with the question of 9/11's impact on digital infrastructure, we explore places where stereotypical accounts of binary relationships between local and global, large and small confront each other in unexpected ways in relation to this site. To do this, we go back to the construction of the Twin Towers in the early 1970s and to the cultural and urban context in which they were embedded.

This reading reveals a much richer context and a more differentiated conception of gigantism than the buildings obtained in and through the terrorist attack. It suggests a more nuanced understanding of the forms of gigantism that the Twin Towers embody and that we explore in relation to the categories of linear, semantic, and latent gigantism introduced in this book. These forms of gigantism need to be regarded on both global and local scales and with both horizontal and vertical perspectives. The excessiveness of their phallic gendering can be challenged by regarding the towers as mediating infrastructures and as containers—as "wombs with a view," as we have seen cultural theorist Sofoulis cunningly describe skyscrapers more broadly.[10]

All of these perspectives coalesce in the Twin Towers and linger in their destruction and in their symbolic afterlife. It is critical to unfold the nuances carefully because they feed into the understanding of latent gigantism, particularly in relation to One World Trade Center, to which we turn in the next two chapters. The Eiffel Tower, which we discuss in the previous chapter, paid a particular tribute to grandiosity by virtue of the tower's rectilinear shape and its use of industrial building techniques, and our reading of it shows how different forms of gigantism intermingle and snap into focus in different ways across the twentieth century. Similarly, the Twin Towers direct our attention to a related but different history of the gigantism of Western urban culture in the twentieth century—one that is no less relevant for understanding gigantism today and that is as much preoccupied with linearity as it is with networked, unstable meanings.

To do so, in this chapter, we first discuss the concept behind the Twin Towers as an urbanistic project and consider foundational ideas of economic growth and progression that were embodied in the World Trade Center project as a symbolic marker of new economic regimes that relied on globally wired financial transactions. We connect these ideas both to global economic developments and to the more local project of the rejuvenation of New York City itself. Moreover, we analyze the architecture and iconic architectural form of the Twin Towers, developing a series of postmodern architectural motifs that the Twin Towers' neo-Gothic architecture embodies. We emphasize their particular form of gigantism—in motifs such as twinning, the double, the cathedral, and phallicism—and also see them as containers intricately linked to the wider urban fabric of Manhattan.

In doing so, we bring in readings of the towers by philosophers who have written on the Twin Towers' architectural and cultural gigantism. This reading historicizes the towers' architecture as much as it historicizes the cultural-theoretical reception of the towers and the cultural issues emerging in the wake of their destruction from prominent postmodern philosophers whose reputations are arguably as grand as the towers' own—Paul Virilio (1932–2018), Jean Baudrillard (1929–2007), Michel de Certeau (1925–1986), and Slavoj Žižek (b. 1949). Although the word *postmodern* means different things in the contexts of philosophy and architecture, we correlate postmodern philosophical critiques that question hierarchies of power—critiques associated with the idea of the postmodern moment as one when the grand narratives of modernity were challenged—with the Twin Towers as a seminal example of postmodern architecture. Reading the towers with and against their reception, we discuss the different forms of gigantism the towers embody. This reading—which "complicates, internationalizes and genders" by bringing the postmodern feminist epistemologies of Donna Haraway (b. 1944) into the discussion (although she does not speak about the towers directly)—suggests a revisiting of the towers that is attentive to a more differentiated account of spatial and temporal scalar relationships and thereby to the forms of gigantism linked to the towers.

The Twin Towers and the Intermingling of Vertical and Horizontal

The 1939 New York World's Fair included an exhibition pavilion called the World Trade Center. Sleek and modernist in style with white walls and cubic forms, the pavilion was dedicated to the concept of "world peace through trade," and the exhibition was run by organizations such as the International Chamber of Commerce.[11] With hindsight, this aspiration on the eve of World War II seems overly optimistic. However, toward the end of the war and in the immediate postwar period, US policy made ideas about reconstructing Europe—and about securing trade- and commerce-driven cross-continental growth and progress—central to the emerging political order of the West. This impetus became an ideological hallmark of the polarized Cold War era. The complexities, contradictions, and naïveté of the visions of progress and progression embodied in this historical trajectory suggest a number of cultural motifs that would stick to the World Trade Center project from the 1939 World's Fair onward.

Although planned in the mid-1960s, the Twin Towers were afflicted by a not dissimilar double bind. The project notably stood in opposition to the mixed uses of industry, commerce, and dwelling that downtown New York had served since Victorian times insofar as the World Trade Center would serve only high finance. The utopian hopes and optimism embodied in the vision of a World Trade Center also tie in with nineteenth-century horizons of expectation of progress and progression: they concern the prowess of industrial culture and the necessity of growth and wealth through market-driven industrial production. In his dedication speech for the towers, chief architect Minoru Yamasaki notably relayed a conception of their architecture as a straightforward form of gigantism—a grand project of cultural salvation through architectural accomplishment: "The World Trade Center is a living symbol of man's dedication to world peace . . . a representation of man's belief in humanity, his need for individual dignity, his beliefs in the cooperation of men, and through cooperation, his ability to find greatness."[12] Yet the allocation of these prospects to a finance-driven economy across the twentieth century indicates the increasing significance of a form of economy that relies on networked mediated relationships, such as that embodied in Yamasaki's World Trade Center.

The networked relationships imply a horizontal expansion that we can observe on different scales, including in the towers' architecture itself. Standing in a tower, you could look out the window and see a precise rendition of where you were yourself standing, a bewildering sameness. Can we see this doubling as a mirror image of the fact that what the towers contained were vehicles of reproduction and transmission? The towers can be seen to reach beyond themselves and tie together other financial centers by being nodes in a global network. Another example is the way the rectangular towers limit the use of glass and give an impression of containment—of preventing something from seeping out. Significant in giving this impression are the dark bands visible on the buildings' exterior, which seem to hold the buildings together like hoops on a barrel. These mark the so-called mechanical floors that were dedicated to electronic and mechanical equipment such as elevator management, heating, ventilation, and air conditioning systems. The embedded technology of the buildings thus functions as a horizontal countermovement to the building's paradigmatic upward energy and at the same time as an essential infrastructure that enables such a tall building to function at all. From the start, the Twin Towers therefore point to both vertical and horizontal forms of gigantism, although some features (such as the antenna, and the waterfront developments at the feet of the towers) were added later.

With its many skyscrapers, New York City is the paradigmatic vertical city of linear, upward-stretching gigantism. In the second half of the twentieth century, New York City was considered to be one of the world's most important cities because of its significance in a postindustrial finance economy rather than because of its previously important position as an industrial center of manufacturing.[13] This leading position in the postindustrial

Looking South on West Broadway toward the towers of the World Trade Center. In DOCUMERICA, the Environmental Protection Agency's photodocumentary project to record changes in the American environment, 1972–1978. Slide, May 1973. © Wil Blanche / Environmental Protection Agency.

The Twin Towers

New York, 1977. © Pierre-Yves Le Bail, 2015. Wikimedia Commons.

Western economy can be seen to have been symbolically cemented while the Twin Towers were going up in the early 1970s, when for roughly two years New York City led the global race to build tall. In 1973, the height marker of the Twin Towers was overtaken by the Sears Tower in Chicago.

The desire to build the tallest building in the world and the fact that this aspiration was bestowed on a world trade center, however, reveal an intense cultural preoccupation with the idea that New York City could become the world's center of gravity by attempting to marshal the world's invisible wealth, represented in the wired spools of trade and financial speculation knotted together in the Twin Towers.[14] The gigantism of the World Trade Center project can therefore be observed as a blatant attempt to (re-)situate New York as a leading metropolis—not just of the United States but of the whole world—in the face of the city's ongoing postindustrial transformations. A concrete height marker of the project's success was the 362-foot radio and transmission tower, installed on the roof of the North Tower (One World Trade Center) in 1978. Yet the antenna on top of the tower enabled operations along the lines of global communication infrastructures and visibly gestured toward gigantic and horizontally distributed

Twin Towers, NYC, circa fall 1993. Scanned from print photos. © Mario Roberto Durán Ortiz, 1993. Wikimedia Commons.

networks. Moreover, the mirroring of the architecture in the South Tower (Two World Trade Center)—and the repetition with a difference in the five smaller buildings (Three, Four, Five, Six, and Seven World Trade Center) that were built between 1981 and 1987—also testifies to a sprouting and reproduction of the structure itself: not only of information through radio and television broadcasting but also spatially as built form.

The gigantic transformational aspirations of the World Trade Center project were underscored locally by the fact that it formed part of a diligently worked-out grand urban scheme that involved all the well-known modernist tropes of regeneration through gentrification and rejuvenation.[15] The World Trade Center as a marker for the success of the financial industry drew people to two neighborhoods that previously had been exclusively industrial and commercial—Tribeca ("triangle below Canal Street") and Battery Park City (along the Hudson River on the lower West Side). Battery Park City was a planned community constructed in the nostalgic image of an urban village and built at the foot of the Twin Towers. It was erected on landfill masses of soil and rock—the by-product of the excavations from several construction projects, including the one that built the underground areas beneath the World Trade Center towers—and it displaced the water and piers of the old river port. Battery Park City is a city within the city, evoking the motif of the double or twinning that we find at the towers' pinnacles. As part of the city's postindustrial development, these downtown areas, which used to be the drivers of New York City's manufacturing economy, now became places where finance sector workers could live and conveniently walk to their offices in the Twin Towers or surrounding buildings. The new situation was symbolized in the collapsing of distance, while the city's industrial heritage lingered as eclectic references.

The World Trade Center development gives form to a vision of a common order of connectivity on multiple scales—from the global reach of the antenna, to the local embedment of the buildings in Manhattan. Battery Park City protrudes into the water as an eclectic historicist montage of urban features. Recalling Benjamin's identification of the Eiffel Tower as an architectural montage, Battery Park City has montage qualities in the way it brings together benches and streetlights that playfully copy and assemble motifs from different periods of New York City's history.[16] As the World Trade Center gathered the agile workers of the world—who were constantly connected to and locked into a gigantic network because trading centers and exchanges are always open somewhere on the globe—Battery Park City embedded the World Trade Center project within a richer urban framework that formed a foundation for the workings of the networked culture of trade that took place out of sight inside the towers.

The historicist and eclectic foundations of the entire urban plan, which encompassed both the World Trade Center and Battery Park City, also point to an understanding of the prowess of Western capitalist culture that is in line with the motifs we explore in the Eiffel Tower. The Eiffel Tower was created for a World's Fair, which can be seen as an extreme manifestation of the interest of late nineteenth-century Western culture in the idea of a cluster or scale model of the richness and superiority of industrial production.[17] In the case of the World Trade Center—where world *trade* as a hub in a large network of exchange had replaced the idea of a world's *fair* as an exhibit of concrete material products—this was a paradoxical manifestation. The materialization of the towers, their vertical gigantism, stood in contrast to the emergence of the new immaterial economy they facilitated, while the urban context was available through the historicist montage of well-known elements collapsed into a friendly urban setting. Notably, however, the place where vertical gigantism and horizontal gigantism met and made themselves present was in the Twin Towers' antenna. Not unlike the Eiffel Tower's intertwinement with radio technology, the antenna of the Twin Towers therefore qualifies as a concrete place where the Twin Towers' architecture and media histories intersect, something that became acutely apparent on September 11, when the antenna's destruction was central to the telecommunication disruptions that occurred.

Challenging the View from Above

We are far from the first to identify the intermingling of vertical and horizontal gigantism in the Twin Towers as a grand architectural and cultural project of the twentieth century. For French philosopher Paul Virilio, the erection of the Twin Towers in the early 1970s signaled a movement from horizontal to vertical, suggesting that the Western frontier was now vertical rather than horizontal—a push for colonization reaching toward the sky. He tellingly ends *Ground Zero*, a book he published shortly after the 9/11 attacks, with the statement that on "September 11, 2001, the Manhattan skyline became the front of the new war."[18]

As part of his writing about urbanism, architecture, and modernity going back to the 1960s, Virilio represents a strong voice against the dominance of verticality and the orthogonal (at right angles), which could be seen in modern architects' predilection for high-rise and tower buildings in the decades after World War II. Virilio thus regards vertically rising skyscrapers as derived from the interest in maintaining surveillance and control over the city. He sees the skyscraper tower as a descendant of the watchtower, a tower of stasis and control.[19] Instead, he promotes the circulation and movement of inclined surfaces, which would affirm a relationship to the movement of the body, privileging the spatiotemporal experience of the situated human body.

This feeds into much larger cultural critiques of skyscrapers and vertical cities as linked to surveillance culture.[20] However, as a counterpoint to verticality, the redemptive potential of horizontal networks and fluidity, so dear to the poststructuralists of Virilio's generation, can also be questioned and should perhaps rather be seen as enabling other forms of control.[21] Indeed, the horizontal as much as the vertical can be regarded as two different manifestations of gigantism, both of which have modes of power and oppression attached to them as well as promises of freedom as vehicles for utopian projections.

Nevertheless, Virilio's position emphasizes that the Twin Towers are inscribed into biopolitical and phenomenological discourses of the twentieth century that wrestle with the dichotomies of verticality and horizontality. The critique of verticality is also central to another towering French philosopher from this period—Michel de Certeau, who is likewise occupied with the decentering of power and who relates this thinking to the relationship between vertical and horizontal figurations in the city. It therefore is significant that Certeau chose the Twin Towers' verticality for his subtle analysis of power relations in the city in a chapter in the much-cited 1980 book *The Practice of Everyday Life*. Let us therefore accompany this Frenchman to the viewing platform atop the South Tower, from where it was possible to look down on Manhattan. His text captures the experience of seeing the city from the viewing platform in poetic terms, foregrounding vision as the central sensory mode of this position and calling attention to the gigantic extremes of New York's urban development, disclosed by this view from above: "*Seeing* Manhattan from the 110th floor of the World Trade Center. . . . A wave of verticals. Its agitation is momentarily arrested by vision. The gigantic mass is immobilized before the eyes. It is transformed into a texturology in which extremes coincide—*extremes* of ambition and degradation, brutal oppositions of races and styles, contrasts between yesterday's building, already transformed into trash cans, and today's urban irruptions that block out its space."[22]

From this position high above street level, Certeau establishes a set of descriptive dichotomies concerning seeing versus not-seeing as positions of power. What is so gigantic about the visual experience from atop the World Trade Center as conveyed by Certeau is the assumption of a superior position overlooking the city. Through this act, Certeau writes, one is "lifted out of the city's grasp": one's "body is no longer clasped by the streets" and instead becomes an "Icarus flying above these waters," "a voyeur."[23] In a similar manner to Virilio, there is a utopianism connected to mobility. But whereas Virilio deems the skyscrapers to be confining structures that weigh

Viewers atop Two World Trade Center observation deck looking north toward mid-Manhattan, June 21, 1984. © Ted Quackenbush, 1984.

down on him oppressively, the skyscrapers for Certeau also embody an upward movement of flight and the ability to assume a liberating position of overview—despite the uncanny resemblances to the colonialist impetus of the cartographer that this position also carries.

What is significant for us here in our cross-reading of the Twin Towers' architecture and the way the towers feature in readings by philosophers such as Certeau is the fact that the viewpoint the tower establishes encompasses an abstraction, whereby the city is turned into an immaterial collection of signs that can be read and decoded. For Certeau, the act of looking involves a transformation of the city "into a *text* that lies before one's eye. It allows one to *read* it, to be a solar Eye, looking down like a god."[24] For Certeau, this view from above is not freed from its political embeddedness. Even from the distanced position atop the World Trade Center, the city reveals itself to the onlooker in all its problematic "extremes," he argues, including social, economic, and cultural inequalities. For Certeau, the city is rich and contradictory—even impregnated with meaning, to use a metaphor that emphasizes the phallic figure implicit in Certeau's description of an "erotics of knowledge" to which his "ecstasy of reading" from atop the World Trade Center belongs. The tall structure becomes for him a *movement* of gigantism, a gigantic instrument or vehicle for assuming a position of overview as much as he is attentive to the illusory qualities of this move.

In contrast to the view from above, however, Certeau does not establish a view from below. Instead, he talks about people at street level as "walkers, wandersmänner, whose bodies follow the thicks and thins of an urban 'text' they write without being able to read it."[25] Despite being blind and thus having no point from which to view the city, these wanderers are not disempowered: they are crucial to producing the "text" that becomes legible only from above. For Certeau, the readers and writers of the city are complementary, and both occupy powerful positions that help to produce the city's wealth of meaning. The way the platform atop the World Trade Center features in his text allows Certeau to contest and also take pleasure in the transformative power of Western culture as imprinted in the built fabric of Manhattan and simultaneously to establish and challenge the intermingling of the vertical and the horizontal gigantism at work.

From the perspective of feminist theory and in a text also published in the 1980s, Donna Haraway broaches the intersection of power, knowledge, and visuality in a related way to Certeau and reminds us that "the god trick"—assuming a position of neutrality and seeing everything from nowhere without being seen—is illusory. As she writes:

> Vision can be good for avoiding binary oppositions. I would like to insist on the embodied nature of all vision and so reclaim the sensory system that has been used to signify a leap out of the marked body and into a conquering gaze from nowhere. This is the gaze that mythically inscribes all the marked bodies, that makes the unmarked category claim the power to see and not be seen, to represent while escaping representation.[26]

Seeing depends on a body that sees, which comes with cultural inscriptions such as gender and ethnicity and which is always conditioned by its own particular posture and ableness. All of this is with us, even if we have the privilege of being able to elevate ourselves to the 110th floor, but should not lead to a romanticizing of the view from below.[27] Haraway's embrace of postmodern epistemologies as outlined in "A Cyborg Manifesto," for example, interlinks this thinking with ideas of networks and horizontal gigantism, emphasizing the networked and relational.

Certeau is certainly aware of the prescriptive factors that Haraway calls attention to, and as art historian Ben Highmore suggests, Certeau, as much as Haraway, is concerned with marking a "way out of indifference" and out of the relativism that lurks in postmodern thought "through epistemological doubt."[28] Haraway emphasizes that "the alternative to relativism is not totalizing and single vision,"[29] and she proposes the partial perspective and situated knowledges as a remedy, proposing to compose them into what she in an early version of "A Cyborg Manifesto" titles an ironic dream of a common language. Reading Haraway in tandem with Certeau's focus on the city's transformation into a text in his account of the view from the platform atop the World Trade Center makes apparent how his approach can be read as a postmodern coda through which the material fabric of the city's buildings appears transparent. It is as if the towers disappear as architectural structures, and with them the conditioning factors embodied by the material context itself become subordinate to the idea of the city as text. In Certeau's writings, the skyscrapers of Manhattan become words or signs—signifiers that stretch up vertically and spread horizontally—that are shrouded in a vocabulary of erotic pleasure that revolves around the ambition of capturing, seeing, and owning the city, even if there is also a deep acknowledgment of the futility of such a project.

Whereas the power figure of the panoptic God's-eye view from above is easy to establish and then critique, the ephemerality of horizontal and relational forms of power is harder to situate in the built environment. When Certeau renders the city as a "text that lies before one's eye," it is through the immaterial features of horizontal gigantism that a spread-out urban plan becomes a gigantically scaled-up text. Certeau identifies this process by considering the Manhattan skyscrapers (including the Twin Towers) as part of an immaterial textual logic of the city. In view of the idea of the postmodern condition as it was described by philosophers at the time, seeing the tower's horizontal gigantism as wholly immaterial, as Certeau does here, is logical because it emphasizes the signs and signification of all the small narratives that fracture the great narratives of modernity. However, this argument does not consider the concrete material and networked relationships that are also part of the towers' gigantism, such as the towers' concrete function as nodal points in a network of mediated relationships pertaining to high-finance trade and communication.

In what follows, we develop what this materiality of networked connections might mean in relation to the towers' controversial postmodern architecture. We thereby turn to the idea of postmodernism not in relation to philosophy but rather in relation to architecture. In architecture, the term *postmodernism* has very particular although contested connotations. Precisely around the time the Twin Towers were being built, historical architectural forms were becoming increasingly significant to architects, an interest that constituted a radical break with the pursuit of the ahistorical formal vocabulary of architectural modernism, which was dominant at the time. The Twin Towers exemplify this burgeoning interest in a historicist architectural vocabulary, not least through the alleged references to Gothic architecture in the buildings' filigree facades.

In philosophy, the term *postmodernism* is often attached to the fracturing of the grand narratives of modernity, but in the 1960s, postmodern architecture emerged from and turned against a supposedly ahistorical modernism and playfully appropriated architectural-historical stylistic elements. When postmodern architecture emerged, it was characterized by an appreciation of ornamental play with signs and signification, and architects often called on a historical repertoire of architectural forms that the modernists' earnest approach to materiality and function had ruled out.[30] Just as postmodernism in philosophy evokes a multiplicity of smaller narratives to confront the grand unifying narratives of modernity and create a new semantic multivalence, so postmodernism in architecture allows (for example, through the use of historical references) a new semantic excess—a gigantism that is less about linear historical time and more about a spatial and temporal structure that expands, collapses, and repositions fragments of the past to create new networks of meanings, much as the city provided Certeau with a wealth of meaning to indulge in and even eroticize.

Is Architecture in or above History?

It should come as no surprise that despite being part of a race to build the tallest human-made building in the world, the Twin Towers looked nothing like the purportedly ahistorical modernist machine architecture of the World Trade Center pavilion at the 1939 New York World's Fair. When they were erected in the early 1970s, the two colossal towers were branded as clean and sleek replacements for a previously disorderly urban site.[31] Reflecting emerging postmodernist tendencies in architecture at the time, however, the Twin Towers were conceived as neo-Gothic superstructures reaching into the sky, like a gigantic double cathedral tower.[32] The self-mirroring effect of the twinning this produced was facilitated by the equally strong motif of the void that marked the space between the two towers. It was as if the towers were both separated and held together by a negation of their own form, an invisible triplet that simultaneously allowed the twin motif to materialize. The constant doubling, even tripling, of the tower's slender but strangely pragmatic form might make onlookers unsure if they were dealing with a cultural mirage or projection, and it created a *Spielraum* for multiple meanings—for semantic gigantism.

The design of the Twin Towers was tied to a different sense of materiality compared with the Eiffel Tower. The boxlike qualities of the architectural form of the Twin Towers made them appear as containers—as "wombs with a view," to use Sofoulis's phrase, which is quoted in the introduction—thereby reversing the phallic imaginaries often attached to this building form. Sofoulis alludes to the way the skyscraper can contain activities and people who are confined to a certain location sheltered by the building but are able to look out.[33] In this light, it makes perfect sense that the tall, slender windows of the buildings were explicitly framed, obviating typical modernist motifs of the panorama window or curtain wall.

The Twin Towers

Two World Trade Center (South Tower) entrance, Liberty Street facade. © Wikimedia Commons.

As containers of people, storage, and wealth, the Twin Towers' architecture engaged the container metaphor in a way that explicitly challenged well-known modernist tropes of transparency.[34] Whereas the filigree structure of the glass-and-steel construction caused the towers to seem to dissolve into air like the Eiffel Tower, it also gave the design a sense of solidity. The Eiffel Tower reduced architecture's constructive elements to molded iron and the free flow of air and made the use of glass superfluous. It stands as a massive air-filled filigree structure reaching into the sky at unprecedented height, while the Twin Towers played a metaphorical game of materiality and mediated transparency that added to the multiplicity of meaning to which they gave rise. Ironically, these wholly material and, in many ways, overly symbolic tower structures—which pointed to motifs of fertility and growth (and are now so evidently imbued with tragic phallicism)—became statements of a new age of networked immateriality in the production of wealth. The Twin Towers were at once concrete vertical superstructures and witnesses to the increasing immateriality of horizontal, networked connections and capital production.

When writing about the architecture of the Twin Towers, the towers' architect, Minoru Yamasaki, emphasized that the design's rationale involved more than engineering principles or even human genius. He denied that the filigree facade had anything in common with Gothic architecture, and he meticulously defended his design choices by using functional arguments. For example, the columns thickened at the base of the tower in order to allow larger window openings in the lobby area. Dismissing readings of this formal gesture as a neo-Gothic ornamental figure, he instead described it in terms of an economy of form, akin to the way that beeswax cells give shape to honeycomb. However, although Yamasaki provides functionalist arguments, something more is at play. Yamasaki quotes the American transcendentalist Ralph Waldo Emerson and his use of nature metaphors, thereby alluding to the possibility that his design points to a dialogue with a transcendental *other* and *more*—a gigantism that itself starts to bear the markers of transcendentalism, as if it stands over and above history.[35] If the towers were constructed to usher in a new world order of wealth and peace for all nations, then they deconstruct the trope of New York as a cathedral city. Their silent transmissions are equivalent to the Gothic cathedral tower, and they both mourn and produce a postindustrial Western city where the nonmanufacturing economy is the main economic driver.

These inherent ambiguities in the design contribute to our view of the Twin Towers as giving architectural form to a culturally transformative point in time. The grand and positivist ambitions of progress and progression, which are central to modernist architectural culture and often seen as its hubris, were met by a burgeoning critique in the 1960s—both within architectural disciplines with the emergence of postmodernism and by cultural theorists such as Virilio and Certeau.

So how can we think of the World Trade Center in relation to the question of whether architecture stands in or above history? To answer this, let us look more closely at what characterizes postmodernism in architecture. Architectural critic and historian Emanuel Petit has proposed that to understand postmodernism stylistically is to misperceive what was a culturally substantial and complex phenomenon. Instead, he suggests that the rhetorical trope of irony is a more appropriate label for postmodernism in architecture.[36] Irony expresses a meaning by using language that usually signifies the opposite (for example, to achieve a humorous or emphatic effect), thereby calling attention to a semantic openness where things are not necessarily what they seem. When described in terms of irony, postmodernism in architecture testifies to the emergence of a form of semantic openness that in Petit's view demarcates the end of high modernism and the onset of something new. Rather than framing postmodernism as a stylistic category affected by the difficult prefix *post*—which always ties postmodernism to the modernism it follows—he instead suggests that we label the entire period of architecture production that spans the early 1970s to 2001 as an age of irony. This periodization encompasses not only the forms of architecture that are typically labeled postmodern but also other -isms that were emerging in this period, including poststructuralism and critical regionalism.[37]

Although we must encourage a questioning of the historicism inherent in the periodizing categories that this argument establishes, it notably foregrounds the Twin Towers. According to Petit, both the beginning and the end of the age of irony coincide with the concrete architectural biography of the World Trade Center. The beginning of the age of irony, Petit writes, was marked by the televised demolition in 1973 of the Wendell O. Pruitt Homes and the William Igoe Apartments (known as the Pruitt-Igoe buildings), a public housing development in St. Louis, Missouri, that became known for its poverty and crime and was demolished some twenty years after it was erected. The destruction of the Pruitt-Igoe buildings has come to represent a systemic collapse of modernism, signifying the failures of modernist architecture and town planning, and it became a sign of architectural and planning disciplines' self-scrutiny during the postwar period. The mediatized atmosphere in which the St. Louis demolition took place and the controlled explosion that fueled it were conscious attempts to create a clean slate on what was considered a defective social housing estate. Thus, a complex interplay between social politics, urban planning, architecture, and changing perceptions of the good life was at work.

The destruction of the Pruitt-Igoe housing estate is related to the Twin Towers in more than one way. Not only were the Twin Towers going up at the same time as the Pruitt-Igoe blocks were falling to the ground, but these two famous and famously destroyed building complexes were designed by the same architect—Minoru Yamasaki. Moreover, Petit argues that the end of the age of irony was marked by the moment the Twin Towers were destroyed on the morning of September 11, 2001.[38] After this point in time comes a new period or age. This historicist interpretation of 9/11 as an

April 1972. The second, widely televised demolition of a Pruitt-Igoe building that followed the March 16 demolition. US Department of Housing and Urban Development. © Pat Bianculli, 2011. Wikimedia Commons.

Twin Towers under construction. © Eric Shaw White. Wikimedia Commons.

event that marked a turning point after which "nothing would be the same again" is something that Petit shares with many other cultural critics.[39] He thereby implies that after 9/11, architects, philosophers, artists, and many others ceased to be attuned to the wider semantic spaces of signification that irony has to offer. Paradoxically, this situates Petit's argument as itself marked by irony. In its preoccupation with the historicist category of the age, it constructs the age of irony as a golden age of what we call semantic gigantism where abundant meaning abounds. This argument buys readily into the philosophical form of postmodernism that Certeau represents when he gives in to the eroticized fantasy that the city is an immaterial collection of signs. It also disregards the way the towers were containers, wombs with a view, and the by no means ironic paradoxes pertaining to issues such as scale, politics, gender, and inequality that arise from this understanding.

Armed with Petit's periodization, nevertheless, we can situate the Twin Towers' erection and destruction quite precisely at not just one but two turning points in architectural history. The argument emphasizes that the Twin Towers can be seen to embody some of modernist architecture's wildest and most naïvely utopian dreams—the race to build tall and the overcoming of the human condition by means of production, progress, and the accumulation of wealth. In this respect, the linear gigantism involved in the towers' architecture should be taken completely at face value. The Twin Towers' architecture is rooted in architectural modernism, but they also summon the burgeoning critique of what modernism had become by the time they were built. Their material form embodies several forms of

The Twin Towers in New York City viewed from below. © Sander Lamme, 1992. Wikimedia.

gigantism—a form of gigantism that grows linearly and also spreads horizontally in the way it opens up for multivalent meanings and networked relationships.

In a stinging critique of the World Trade Center project published in 1966, the *New York Times* architecture critic Ada Louise Huxtable addresses the impact of its gigantism and its future posthistory: "Who's afraid of the big, bad buildings? Everyone, because there are so many things about gigantism that we just don't know. The gamble of triumph or tragedy at this scale—and ultimately it is a gamble—demands an extraordinary payoff. The trade center towers could be the start of a new skyscraper age or the biggest tombstones in the world."[40] Huxtable situates the buildings in the tension between material and immaterial, ground and sky, symbolism and profit, in a way that problematizes Yamasaki's transcendentalism. This view is connected to the towers' anachronistic air, which Huxtable pointed to: the rectilinear form of their vertical gigantism was already outdated at the time of their erection, when much more fluid modes of control and internalized disciplinary regimes had long replaced the idea of the looming watchtower.[41]

Nevertheless, the uncanniness of Huxtable's image of the Twin Towers as gigantic tombstones is biting and at the same time inscribes the buildings in an explicitly Gothic framework. This emphasizes the towers' architectural historicism in a way that opens a register of irony that is more indirect than Petit's terminology implies. It questions the temporality of progress and progression while still relying on it. Rather than indulging in the widened frame of reference that the question of modernism's fantasy of progress and progression makes possible, Yamasaki distances himself from an understanding of the widened space of signification that irony offers. We can argue that the Twin Towers were not entirely postmodern in Petit's sense. Could this explain why they met with such a spirited critique from Huxtable, who was more attentive to the complexities of the towers' historicist underpinnings than even arguably the architect himself? The strangely evasive character of the towers' gigantism (as commented on in the prologue to this chapter) remains a marker of the ambiguities imprinted in their design, their conflicting history, and the latency of their continued looming significance.

The coexistence of different forms of gigantism calls on conflicting understandings of time in the Twin Towers project and illuminates a contradiction. It is as if the design of the towers is simultaneously set *in* and *above* history—or that they are both postmodernist and modernist at the same time. As a gigantic twin womb with a view, the towers simultaneously fostered the architectural vocabulary of which they were part *and* provided a viewpoint on that vocabulary. Thus, they became a monument to a time that had already passed, not unlike how Barthes viewed the Eiffel Tower as a monument to an already eroded symbolism. It should therefore come as no surprise that the Twin Towers have been described as "an already decaying monument to a form of society that has itself already been eroded and displaced by new more problematic forms," to quote political scientist Julian Reid writing on the towers.[42]

The Twin Towers' architecture thus simultaneously reflected, produced, and in a curious way anachronistically repelled the temporal conditions that materialized in, around, and with them at the time of their erection. If the buildings were marked by ambiguity, their tragic ending itself put an end to that ambiguity. But what do the scale and brutality of that ending do to the complexity of the gigantism at work here, and how has it been channeled by the philosophers who saw these buildings as vertical markers of modern culture's mishaps?

Towering, Falling

The gigantism of the Twin Towers made them a target of monumental proportions—in physical size as much as symbolic value—on September 11, 2001. The symbolic value of the gigantic event itself brings the different forms of gigantism that we have uncovered into play once more, at the same time as it begins to reveal additional temporal contradictions in which both past and future are folded into the present. This line of thought has been pursued by the Slovenian philosopher Slavoj Žižek, who was quick to respond to September 11 and who published the essay "Welcome to the Desert of the Real!" shortly after the attack. Here, Žižek argues that we had already been prepared for this catastrophic event, and even anticipated it, through numerous Hollywood films. For him, September 11 represented a crude awakening to the realities of American foreign policy:

> It is precisely now, when we are dealing with the raw Real of a catastrophe, that we should bear in mind the ideological and fantasmatic coordinates which determine its perception. If there is any symbolism in the collapse of the WTC towers,

it is not so much the old-fashioned notion of the "center of financial capitalism," but, rather, the notion that the two WTC towers stood for the center of the *virtual* capitalism, of financial speculations disconnected from the sphere of material production. The shattering impact of the bombings can be accounted for only against the background of the borderline which today separates the digitalized First World from the Third World's "desert of the Real."[43]

Žižek pinpoints how the material and immaterial, vertical and horizontal forms of gigantism that we have identified as characterizing the architecture of the Twin Towers are also at the heart of their destruction. Significantly, he evokes dream imagery, which we also saw both Walter Benjamin and Roland Barthes remark on with regard to the Eiffel Tower in chapter 1. Here, however, the dream—in the form of virtuality—is contrasted with the real as a gigantic ruinous desert, through a Derrida quotation used in the 1999 film *The Matrix*.

The aftermath of the attack is thus certainly no less monumental than the attack itself and is possibly even more so, according to Žižek. He reverses the imagery so that the virtual and digitalized come to describe a contained, even confining world from which it is possible to awaken, whereas the real vastness lies outside this bubble. In this way, Žižek speaks to a different kind of horizontality than the global telecommunication infrastructures we have dealt with so far. He brings attention to the discourse of a digital divide between what he calls the first and third worlds, an inequality with regard to the economy and to access to technologies. Although Žižek might be right that the attacks gave way to something else, this has not proved to be a vastness beyond digital infrastructures, which have only increased their reach across the globe. Yet Žižek's remarks show us the degree to which September 11 as an event is ingrained in the vocabulary of the gigantic and intertwines intricate notions of materiality, immateriality, verticality, and horizontality. To comprehend the gigantism of the attack itself (and the gigantism of what followed, to which we turn in the next chapter), we need to look more closely at the relationship between the destruction in architectural and mediatic terms and the ways the towers came to be understood after their collapse.

The destruction of the World Trade Center towers happened on the morning of September 11, 2001, when two passenger planes were flown into the towers by hijackers. One plane was piloted by Mohamed Atta, who had studied engineering and architecture in Cairo and Hamburg and who

Smoke rises from the site of the World Trade Center, Tuesday, September 11, 2001. © The US National Archives, George W. Bush Library.

was known to be highly critical of Western culture and of skyscrapers as a cultural form.[44] Within hours of the attack, both of the Twin Towers—as well as the adjacent Seven World Trade Center tower—collapsed, emitting rubble, smoke, and dust that covered Manhattan in a ghostly white cloak. This material distribution of dust gave it an uncanny media quality on a very large scale as the material debris became a communicative layer and covered the city as physical testimony to the event.

Shortly after the attacks, French philosopher Jean Baudrillard—who was from the same generation as Virilio and Certeau and who in the 1960s had already made a strong critique of the biopolitical implications of vertical gigantism—described the attack as having been provoked by Western culture itself, a reaction that he metaphorically likened to bulimic overeating: "It is probable that the terrorists had not foreseen the collapse of the Twin Towers (any more than had the experts!), a collapse which, much more than the attack on the Pentagon, had the greatest symbolic impact. The symbolic collapse of a whole system came about by an unpredictable complicity, as though the towers, by collapsing on their own, by vomiting suicide, had joined in to round off the event. In a sense, the entire system, by its internal fragility, lent the initial action a helping hand."[45] Baudrillard's phrasing aligns with other critical readings of the attack that did not simply regard it as an assault on liberal societies from a perceived outsider but rather situated it in a trajectory of resistance to the discipline and control at work within such societies.[46] If we bracket the political underpinnings of the argument for a moment and focus on its ontological implications, we see that Baudrillard assigns the meaning-endowing function of the event to the towers themselves. It is as if the falling towers were agents capable of political action, supposedly "joining in the event" with their collapse. In this way, Baudrillard conflates and reverses human and material, action and impact, by anthropomorphizing the towers in an even more radical way than Barthes did with the Eiffel Tower.

Another voice from the same generation, the German avant-garde musician Karlheinz Stockhausen (1928–2007), took this argument a step further by drawing on imagery of the Romantic sublime in a provocative coupling of affective response and aesthetic effect. In a German radio interview on September 19, 2001, Stockhausen argued:

> What happened there, is—now you must all reset your brain—the greatest artwork ever. That spirits accomplish in one act something that in music we could not dream of; that people rehearse like crazy for ten years, totally fanatically for one concert and then die. That is the greatest artwork for the whole cosmos. . . . In comparison, we as composers are nothing. Imagine that I could now create an artwork and all of you would not only be amazed, but you would drop down on the spot, you would be dead and reborn, because it is simply too insane. That is what many artists also try to do, to go beyond the limit of what is thinkable and possible, so that we wake up, so that we open ourselves for another world.[47]

Stockhausen describes the attack as a work of art whose sublimity is out of reach for the artist. That the final effect of a symbolic death could in fact be death itself is for Stockhausen the ultimate yet unattainable effect of art. Although this radical aestheticization represents a certain aloofness and links up with Romantic motifs that haunt twentieth-century radical politics and art alike in uncanny ways, we may note a peculiar and perhaps even more uncanny resemblance to Yamasaki's transcendentalism. As noted earlier, the World Trade Center towers were not the first of Yamasaki's buildings to collapse. However, the possibility of the clean slate or tabula rasa was for Yamasaki, as for so many architects and avant-garde artists of the twentieth century, above all an opportunity for renewal, for change, and for new life.

In the aftermath of the destruction of the World Trade Center, moreover, many planners saw any rebuilding as an opportunity to connect the old downtown more successfully to Battery Park City.[48] It thus became the occasion for a renewal that many had been looking for, not unlike the destruction of the Pruitt-Igoe buildings before it. However, there is a notion of sublimity at stake in these responses that holds its own form of gigantism. Such sublimity is indebted to the idea of the destructive event and the ultimate disruption.[49] It is part and parcel of the narratives of grandness of these buildings: is a sublime catastrophe of gigantic proportions enabled when you build the tallest building in the world? The project's combination of vertical ascension and horizontal repetition also led Jean Baudrillard (in a conversation with Jean Nouvel published in French a year before the 9/11 attacks) to describe the towers, in a highly dystopian spirit, as "the end of the city": "The World Trade Center expresses the spirit of New York City in its most radical form: verticality. The towers are like two perforated strips. They are the city itself and, at the same time, the vehicle by means of which the city as a historical and symbolic form has been liquidated—repetition, cloning. The twin towers are clones of each other. It's the end of the city."[50]

So far, however, the announcement of the end of the city has held no truer than Žižek's announcement of the desert of the real. To the contrary, we see vertical cities shooting up across the globe, more digitally wired and "smarter" than ever before. The strong symbolism of the motif of falling buildings as an act of cleansing foreshadows the intentions behind the destruction of the World Trade Center orchestrated by terrorist forces. The vision of the destruction of architecture as emblematic of the establishment of a new order of reality is also arguably present in the alleged ISIS video of the falling Eiffel Tower to which we referred in the previous chapter. We are grappling here with a truly modern trope that links destructive events, even catastrophes, with the possibility of turning the cultural tide.

The architectural history of the Twin Towers is tightly interwoven with their media history. To consider the mediation of the buildings' destruction, we can return to Chun's analysis of the attunement of networked digital media to a perpetual state of crisis that prompts a real-time response, which may then cause a sense of inadequacy if no such response is given. In the prologue to this chapter, we saw how this anticipation of a real-time response, which we take for granted today, afflicts one's memory of hearing about the fall of the towers. Yet 9/11 happened a few years before social media and the notion of Web 2.0 gained momentum. As we saw at the beginning of this chapter, the telecommunication infrastructure looked significantly different less than twenty years ago when Google advised users to watch TV or turn on their radios when traffic soared on September 11, a situation that seems improbable today. The destruction of the towers is thus situated in the interim between the televising of catastrophe and the permanent state of crisis that governs social media's compulsive stream of breaking news and clickbait begging to go viral.

Stockhausen represents a belief in art's ability to jolt us out of the habitual, and he therefore embraces 9/11 as a catastrophe that accomplishes exactly that. However, we may say that it has become more difficult to imagine anything that might startle media consumers like us enough to have that effect at a time when crises are habitually consumed and distributed online in the ways seen with the online responses to the terror attacks on Paris in the previous chapter. The way the Twin Towers' architecture is ingrained in media history emphasizes their position not only as containers pregnant with information, people, and wealth but also as viewing platforms from which to observe the historical time of which they were a part. This observation reinstates the Certeauesque conundrum of how to grasp the city in a way that does justice to the semantic gigantism at work.

What we find here is an unresolved and internally contradictory form of gigantism that expands in a seemingly endless play of meaning—thereby questioning its own ability to be monumental—and at the same time also reaffirms linearity insofar as the architecture is stubbornly growing so excessively high in order to become the highest tower in the world. This inherently unsettled form of gigantism is what lingers and contributes to the towers' latency that influences, as we shall see in the next chapter, the form of gigantism that imprints what came after the Twin Towers on the World Trade Center site. In "A Cyborg Manifesto," Haraway notes: "For salamanders, regeneration after injury, such as the loss of a limb, involves regrowth of structure and restoration of function with the constant possibility of twinning or other odd topographical productions at the site of former injury. The regrown limb can be monstrous, duplicated, potent. We have all been injured, profoundly. We require regeneration, not rebirth, and the possibilities for our reconstitution include the utopian dream of the hope for a monstrous world without gender."[51] The question to which we now turn is the particular form of gigantism regrowth after injury has taken at the Ground Zero site, but also the potentially monstrous utopian dreams that might stick with us today.

Maria Finn, *Unfinished #17*, pencil on paper, 29 × 42 cm, 2018. © Maria Finn.

3 The One World Observatory: Caught between Vertical and Horizontal Gigantism

Prologue: Gigantic Buildings

New York, April 2016

The night before I left Copenhagen to meet Kristin in New York City, I dreamed about a future architectural development of Copenhagen. Despite the sprouting of buildings in the twelve-story range in recent years, Copenhageners are known for their ambivalence toward tall buildings. Nonetheless, in the dreamscape, a skyline of fully high-rise architecture was added to the city in a truly spectacular manner. As if I were on an airplane, cruising at low height and looking down on the city, in my dream, the skyline that unfolded before my startled inner gaze was more akin to distant cities like Dubai or Shanghai, collapsing visions of the future onto the Copenhagen I know so well.

Approaching the city from the east and seeing very little of Copenhagen proper, I first recognized Spanish architect Santiago Calatrava's actual fifty-story tower, the Turning Torso. The tallest building in Scandinavia, this tower was built in 2005 close to the waterfront of Malmö, Copenhagen's nearest neighbor on the Swedish coast. In my dream, I apprehended Copenhagen in the wide regional sense implied in the 1990s Danish-Swedish political project to create a single urban region across the Øresund Strait. The opening of the Øresund Bridge in 2000 was the most visible infrastructural project supporting this vision, and along with it came a railway line ensuring an easy commute across the strait. However, with the surge of refugees arriving in Sweden in 2015, fifty-eight years of open borders between Denmark and Sweden were terminated, and commuters were suddenly prompted for IDs and passports.

Nonetheless, the vision of regional development and joint prosperity for the neighboring countries extends beyond the building of new infrastructure. It includes large swaths of commercial, institutional, and residential architectural development, not least in the southern part of Copenhagen, where an entirely new urban area called Ørestad has been built from scratch since the early 2000s. Ørestad is known for a series of fairly tall and architecturally spectacular buildings, several of which have become world famous, such as the 8 House by

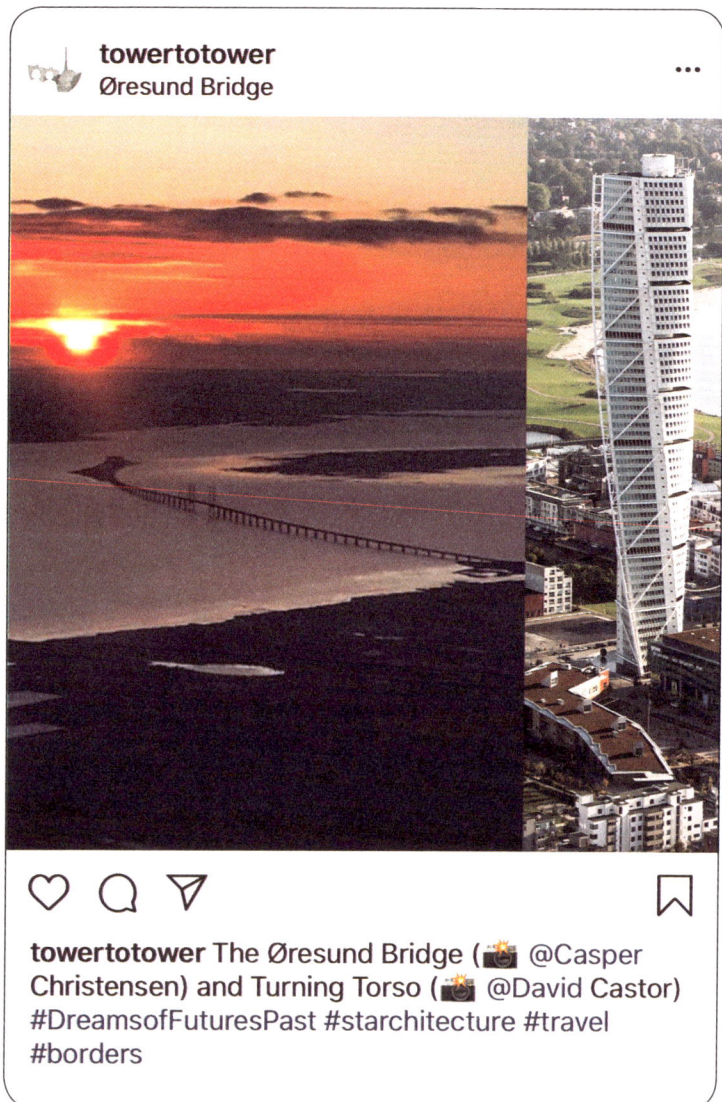

Danish architecture firm Bjarke Ingels Group (BIG). Parallel developments have taken place in the smaller city of Malmö, where Calatrava's landmark tower, so vividly present in my dream, is among the most significant features. The tower was created in the gigantic anthropomorphic image of Calatrava himself, projecting a likeness of the architect's body as he gently twists his torso toward his own image in a mirror behind him.

However, the visual power of Calatrava's megalomaniac, notoriously phallic, and very real tower fades in comparison with the lineup of buildings that followed in my dream. As I continued my airborne approach toward Copenhagen, gliding above the water, a series of spectacular new glass-and-steel high-rise buildings appeared on a straight row of islands connected by a superhighway across the Øresund Strait. The footprints of the buildings covered most of these little islands, so that it looked as if the buildings had their feet in the water. The calm blue surface was reflected in the glass fronts of expressive yet geometric buildings with no facades, like the black-box skyscraper one might find in many cities all over the globe today. The climax of this impressive skyline was a gigantic glass-and-steel boxed-in structure in the shape of the Eiffel Tower. So here it was, in my dream—many times the size of the original and emitting sparks of glitter. This was the Eiffel Tower's dramatic and fairy-like twin sister, a ballerina queen on a cocaine high. Highly spectacular yet only diffusely present. I could not tear my eyes from the structure, infatuated by the very sight of it.

Then a voice invaded my dream: "Mom, are you leaving today?" spoken by an anxious and still half asleep five-year-old. The words propelled me into the early-morning atmosphere of my dimly lit bedroom, but the image of that fantastic skyline lingered on. A few hours later, when I was on an actual airplane leaving Copenhagen, I looked down on the city and the water from above—just checking in case what I had seen in my dream had magically appeared overnight.

In this chapter, we turn to One World Trade Center, a building that looks as black-boxed as any of the skyscrapers from my dream. We consider this building as we experienced it when we first visited in April 2016 (the trip that commenced with the dreamscape above)—as simultaneously tower, transmitter, and tourist site, as an architectural object and mediated spectacle, and as marked by latent gigantism.

—HS

The Sky's the Limit

Completed in 2014, the gigantic structure of One World Trade Center is visible from a surprising number of places on Manhattan. You notice from far afield its reflective surface, sharp edges, and sleek if somewhat plump form, which narrows ever so slightly before culminating in the antenna at the tip of the building. Yet One World Trade Center is noticeable in a different way than the Eiffel Tower, which stood out visually to such a degree that Guy de Maupassant allegedly chose to have lunch there in order to avoid seeing it. Unlike the Eiffel Tower, One World Trade Center does not seem to stand out from the greater urban landscape by being overtly avant-garde, a statement of newness, or a pinnacle of what engineering construction can accomplish. Nor does it raise itself above Manhattan as a mirroring monument to the transition from industrial culture to information society, as the Twin Towers did.

Instead, it stands out by its relative anonymity, less singular and more commonplace in its mimicking of so many other corporate skyscrapers in Manhattan and elsewhere. The building makes its mark not through innovative or expressive design but through sheer size. Surpassing the Empire State Building (just as the Twin Towers did from 1970 until 2001) as the tallest structure in New York City, with its height of 1,776 feet or 541 meters, One World Trade Center takes part in the same race to the sky that its many modern predecessors joined in. However, at the same time, it suggests a natural ending to that upward race. Measured in American feet, its height equals 1,776 feet, which is the year of the country's independence, a symbolic number that gestures toward the heart of the gigantism involved in the "American dream."

The reference to the year 1776 as a design gesture measured in feet indicates a significant engagement with history and temporality but also with hope and finitude. In One World Trade Center, this engagement differs in significant ways from the other pieces of architecture discussed in this book so far. Rather than overtaking other buildings in the sense of being a frontrunner or by playfully pointing toward the past, present, and future as the Twin Towers did, One World Trade Center is a mainstream colossus that blends calmly into its surroundings, leaving only its size and its sleek, impenetrable surface to mark the many interests at work on the site and the compromises involved in erecting a building there. The building continues

The One World Observatory

One World Trade Center, Lower Manhattan skyline, October 2014. © Phil Dolby, 2014. Wikimedia Commons.

urban planning's modern race to build tall by erecting ever-taller skyscrapers and the implications of linear gigantism. But at the same time, as we argue, both the antenna on the roof of the building and the mediatization in which the experience of the building is enfolded point to vertical, horizontal, semantic, as well as latent forms of gigantism.

With this recently constructed building, our architectural analysis therefore finally catches up with the conundrum that is the main driver of this book and that we broach with regard to the digital incarnations of the Eiffel Tower in chapter 1. That analysis concerns the question of what is at work in gigantism today, when the connotations hitherto ascribed to gigantism (linear progression, frontier advancement, semantic widening) are being challenged by the planet's multiple and equally gigantic crises: Why build an excessively tall building today, when by many accounts it is unsustainable? Like the other towers in this book, we consider how One World Trade Center is simultaneously *both* a phallic superstructure from which wireless signals travel even further into an abstracted informational space *and* a container of people and information that provides a viewing platform onto the historical conditions out of which it has emerged. Our reading of One

World Trade Center therefore explores points of intersection between the building's architecture and the digitally rendered cultural mediatizations that envelop it in order to come to grips with the latent gigantism at work here and the way it amounts to more than the eye can see.

The destruction of the Twin Towers made way for something that was still unknown that September morning in 2001. As discussed in the previous chapter, architectural historian Emmanuel Petit interprets the moment when the Twin Towers collapsed as a wake-up call, a cultural summation whose meaning was impossible to miss. For him, it marked the end of the age of irony and the onset of something new. This echoes the voices of a range of cultural commentators on 9/11, including, as we have seen, intellectuals such as the Slovenian philosopher Slavoj Žižek and the German avant-garde composer Karlheinz Stockhausen, who both saw the event as a turning of the cultural tide.[1]

This understanding is present in the conceptualization of the destroyed World Trade Center site as a Ground Zero. The name echoes the idea of the end of World War II as a *Stunde Null*, Hour Zero. This phrase famously marked the end of the war as an absolute break with the past and a radical new beginning for Germany as the country struggled to move on from war, Nazism, and the holocaust. Naming the site of the former World Trade Center as Ground Zero foregrounded the gigantic impact and continued reverberations of the attack. It is not our intention to diminish the cultural and political impact of the event or the impact on the people and families affected. However, there is something symbolically monstrous and overtly gigantic about likening September 11 to World War II.[2] As for the Twin Towers themselves, even after they had collapsed, they continued to linger on in a wealth of cultural imaginaries and images of New York City. And they would continue to loom over the discussions of what should go up on this site as it became ready for architectural renewal.

This chapter both uses and challenges the idea of 9/11 as a marker of change and considers the architectural culture to which One World Trade Center belongs—the building that was eventually built on the Ground Zero site after more than ten years of discussion over its design. The Twin Towers, as we discuss in the previous chapter, were marked by an intermingling of vertical and horizontal gigantism that leaned toward postmodern ideas about semantic widening and the play of signifiers and architectural forms, relating to what Petit calls an age of irony. But if the new One World Trade

Center is markedly different, as we propose, what kinds of gigantism greet us in this tower, and what is the relationship between them?

There is a contradiction between Petit's philosophy of history, which sees time's progression as a succession of ages, and our argument, which challenges linear understandings of time by conceiving of a broad present in which highly distributed spatial phenomena coalesce with a sense of simultaneity and temporal overlay. However, if we accept this contraction for a moment, allowing both ideas to linger side by side, it follows that after the age of irony and the semantic excess, complexity, ambiguity, and multivalence that irony implies, there came an age marked by a certain straightforwardness—even though this straightforwardness may be seen to divert attention from, or even shield, more difficult and contingent phenomena. This means that One World Trade Center alerts us to a form of signification that is neither modern (in the manner of the Eiffel Tower, concerned with advancement and progression) nor postmodern (enabling multiple coexisting meanings, in the way envisioned by the late twentieth-century thinkers discussed in the previous chapter). Rather, we argue that One World Trade Center embodies a form of signification in which the mediated and material levels of the building intermingle in a series of temporal overlays that are configured in an intricate relationship between signaling security, selling cultural meaning to visitors as consumers, and offering a highly curated space for particular forms of participatory engagement—forms that facilitate both security and consumerist paradigms and that operate through a sense of latency.

In this chapter, we first discuss the alleged straightforwardness of the architecture of One World Trade Center as a species of the urban planning paradigm that the Swiss sociologist Christian Schmid calls the *new metropolitan mainstream*. This discussion helps us to unpack the wider urbanist ideas implied in the design. We consider the building's outer appearance and effect. Entering through the tourist entrance, we focus on the concrete experience of the tower as a tourist site and the integration of mediated views into this experience. We focus on the part of the building that is available to tourists rather than on the corporate office culture, which is largely invisible and unavailable to the public (although it takes up the majority of floor space of the building) because we regard the highly mediatized tourist experience as a key marker of the building's sleek rapport with the new metropolitan mainstream. We thus explore the way the building makes use of what we could call a trick of the contemporary urban experience economy: a tiny fraction of the building is open to the public (or that part of the public that is able to pay a significant entrance fee) and is heavily branded and swathed in mediatized experiences that attune the visitor to a particular set of narratives about the building.

Apart from being an office building, One World Trade Center is simultaneously tower, transmitter, and tourist site. It is architecture, media, and mediation. Exploring the building's architecture in this triple sense reveals particular meanings embodied in the building's design, the cultural context to which it refers, and the form of communication it performs. By contrast, the next chapter, chapter 4, considers the contribution of this visual order to the meaning of the public memorial space in front of the tower on the Ground Zero site. In this way, we circle in on the kind of meaning possible at this site—a context where the latently present, mundane, and invisible office culture of the gigantic skyscraper intermingles with the narrative of the building as a publicly available symbolic site, a building that towers over and provides a platform for seeing the surrounding city from an elevated position.

Because the One World Trade Center design is simultaneously straightforward and indirect in its gigantism, our reading struggles with what we consider to be a parallelism of effects. One World Trade Center brings our attention to ontological slippages as both a concrete piece of architecture and a media platform. Our identification of these parallels touches on the core characteristics of the latent gigantism we associate with the building. At the same time, we also recognize that our reading builds an entrapment: is it possible to negotiate this ontological slippage theoretically without reproducing it conceptually? This question is connected to questions about how humans, technology, and materialities interact and are interdependent, as well as how the categories of past, present, and future permeate the site. We broach these concerns in our final discussion in this chapter, where we consider the epistemological problems behind struggles over the One World Trade Center design in light of the simultaneously expanded and flattened temporal notions that accompany latent gigantism—not just in the spatial, architectural, and mediatized environment in and around One World Trade Center but also in relation to the theoretical concepts we have at hand to understand the character and effect of this latent gigantism.

One World Trade Center and the New Metropolitan Mainstream

The negotiations that led to One World Trade Center's current design have been well documented by both researchers and public media. What interests us here is how the initial design proposal for the building in 2003 was rejected, what proposal replaced it, and to what extent we can see contrasting forms of gigantism in the two proposals.[3] In 2003, when Polish American architect Daniel Libeskind won the architectural competition for the World Trade Center site, it seemed a striking fit. Who else could take on the responsibility of the reconstruction of that site? The Jewish Museum Berlin, designed by Studio Libeskind, had opened in 2001, vaulting Libeskind to fame as one of the 2000s' most significant starchitects—a catchy if ill-defined term that elevates unique creative minds to heights as towering as their expressive architecture. Although he was the subject of critique inside and outside the architectural profession,[4] Libeskind was widely regarded as an architect with the creative and philosophical capacity to respond to the most scarred sites of human culture. The idiosyncratic and highly personalized narrative embedded in the Ground Zero design is in dialogue with characteristic forms of postmodernism, linking it with the architectural paradigm of the Twin Towers more intimately than may at first appear from comparisons between Libeskind's expressive forms, which tap into emotional registers, and Yamasaki's subdued transcendentalism. In line with the previous urban plan for the World Trade Center site, moreover, Libeskind's proposal envisioned the site as a cluster of buildings. Yet it did so in a way that would leave open the exact location where the Twin Towers had stood after the rubble was removed. The towers' footprints would seemingly remain untouched as authentic markers, respectful of the tragedy that had taken place here and exposing the very bedrock of Manhattan as a symbol of an open wound that would create a place for individual and collective commemoration.

Moving through a number of symbolic articulations that are present as dialectics between light and darkness, sky and earth, death and renewal, Libeskind's design, poignantly titled the Freedom Tower, envisioned a single new tower spiraling 1,776 feet upward, echoing the year of American independence and mirroring the upward-reaching arm of the Statue of Liberty. Moreover, in a pseudo-cultic gesture, on September 11 every year the site would be lit by two glimpses of sunlight at the exact times when

Master plan sketch. © Studio Daniel Libeskind.

the two towers of the World Trade Center were hit by planes, a symbolic reenactment of the moments of their destruction that Libeskind called the Wedge of Light.

Libeskind's ability to articulate a level where architecture embodies the deepest ethical concerns, suggesting a place where a collective confrontation with shared cultural trauma and emotions can occur, was crucial to his winning this project. He visualized in tangible form what seemed impossible to represent, and he intended the built fabric to open a spectrum of symbolic meaning. This reminds us of the semantic widening that characterized the Twin Towers' postmodern architecture, although Libeskind is notoriously vocal in his storytelling about his buildings, where much of the symbolic pathos is fed through dialogues between the architect's speeches and texts about the buildings and the actual built form.

Moreover, the ragged and idiosyncratic shapes of Libeskind's designs can be associated with the increasing use of digital technologies in the process of form-finding in architecture since the 1990s. Notably, the argument for the significance of digital technologies was put forward in 2002 in an updated volume on postmodernism by the most significant writer on such architecture, Charles Jencks, called *The New Paradigm in Architecture: The Language of Postmodernism*. The book features Libeskind's Jewish Museum on the front cover.[5] However, this expressive aspect of the design and its elusive, sophisticated references to multiple symbolic meanings were also what became most problematic about the project. It was as if this form of architectural thinking lost its currency at the very moment when Jencks announced it as a new paradigm, and its abilities to digitally mold wild fantasies became operationalized in more economic and less creative registers.

In Manhattan, the tower that was eventually built on the Ground Zero site was reworked to the point where very little of Libeskind's design proposal was ultimately realized. The difference between these paradigms is summarized when, in a newspaper article, a statement by Daniel Libeskind emphasizing dramatic effect and emotional resonance (what he calls emotional geometry) over quantitative measures is juxtaposed with the interests of one of the nonprofit organizations representing the victims, survivors, and others with a vested interest in the representation of 9/11:

> "In a perfect world, it might have been preferable for the pools to coincide precisely with the tower footprints," said Daniel Libeskind, the master planner of the trade center site. "But their emotional geometry and integrity are more important. They are at the epicenters of where the towers stood, and much like the twin beams of light, though not exact replicas, they accurately and movingly reflect the power of absence." Anthony Gardner, executive director of the World Trade Center United Family Group and a leading preservation advocate at ground zero, does not agree. He said the failure to replicate the towers' dimensions and their distinctive cutaway corners "minimizes the enormity of the buildings themselves, the scale of the loss and what was taken."[6]

The opposition emerging here notably exemplifies in architectural terms Heidegger's identification of gigantism, which we quote in the introduction, as transcending what is calculable. What is at stake here is not only a quantitative measure but something that slips into a qualitative regime and concerns questions of feelings and affect.

The One World Trade Center we encounter at the site today was designed by a completely different architect—David Childs of the architecture firm Skidmore, Owings, and Merrill. Libeskind instead became mainly the master planner of the site rather than the architect of individual buildings. The One World Trade Center tower is part of the overall topography of the Ground Zero memorial site. As Libeskind's original design suggests, it was moved away from the original World Trade Center site, which was left vacant to make space for a memorial marking the shapes of the Twin Towers (the design of which we discuss in the next chapter). One World Trade Center hovers beside the memorial site as a thick, completely opaque office tower, shielded by the impenetrable reflective glass facade. Apart from the tip of the antenna, the building makes no evident reference to Libeskind's imagined original. Likewise, the most evident reference to the Twin Towers is the building's height, which (minus the antenna) is approximately equal to that of the roof of the old North Tower. One World Trade Center is clearly gigantic (it is one of the tallest buildings in the world), but its gigantism is different and much more subdued than that envisioned by Libeskind, which foregrounded architecture's rootedness in history and culture and embraced architecture's capability for expressing multivalent meanings.

At first glance, One World Trade Center presents itself as a sleek, gigantic office tower that is in tune with the smooth global style expected of such buildings in prominent urban sites. Arguably, it does not present itself as a particularly remarkable building, if we disregard the site it occupies and the fact that it is of gigantic proportions. Indeed, One World Trade Center is the tallest freestanding structure not only in New York City but in the United States, and it is the tallest building in the Western hemisphere (superseded by the CN Tower in Toronto as the tallest freestanding structure, although the latter is not a building).[7]

We argue that the building's sleek skyscraper aesthetic, rather than adhering to the postmodern paradigm proposed by Charles Jencks, resonates with what Christian Schmid has called the new metropolitan mainstream.[8] The *new* in *new metropolitan mainstream* does not indicate that there

One World Trade Center, New York. © Sergei Gussev, 2016. Wikimedia Commons.

is a succession of one thing after another but that the urban metropolitan forms of industrial culture, dating primarily from the late nineteenth and early twentieth centuries, are taken up again. The cultural and architectural figurations employed in the new metropolitan mainstream draw on imagery that can be associated with an early metropolitan cultural elite. It is preoccupied with the unique, the exceptional, and the avant-garde, and it engages with these concepts by developing expressive forms of architecture. The process of taking largely historical avant-garde or industrial formal paradigms and making them generally pleasing and mainstream places the new metropolitan mainstream on a par with other appropriations of avant-garde techniques from the early twentieth century, which became institutionalized throughout the century (such as the montage technique we discuss in chapter 1).

The urban plan of Battery Park City we discuss in the previous chapter implied a montage quality in dialogue with the twentieth-century avant-garde aesthetics that Benjamin described, but the appropriation in this context bears the mark of the way in which digital technologies have made previously laborious work (such as montage and other cut-up forms) available to more people by the click of a button. This can be seen in skillful appropriations in fan art that is often of equally high quality as the original on which it is modeled.

As we discuss in relation to the prefix *post* in *postmodernism* in the previous chapter, however, prefixes such as *post* or *neo* are temporal notions and inevitably also concern questions in the philosophy of history. In this respect, we argue, the idea of the new metropolitan mainstream can be seen as a suggestion to explain what comes after postmodernism or what we in the previous chapter refer to as the age of irony. This is not as a desert of the real, in Žižek's words, but rather as the recent embracement in architecture and urban planning of what was once avant-garde architectural vocabulary to an extent that it becomes mainstream, even habitual and unnoticeable.

In the new metropolitan mainstream, the idea of the metropolitan refers to different forms and figurations associated with the urban culture of the late nineteenth and early twentieth centuries.[9] The glass-clad skyscraper is one such form. It points to a narrative of progressiveness, to the centralization of wealth and production, and thus to vertical gigantism as part of a narrative that aimed to set metropolitan centers apart from what was conceived as a more regressive hinterland. It marks how the big metropolitan centers of the late nineteenth century, such as Paris and New York City, would raise themselves as world centers of production and power in an industrial economy. However, as Schmid argues, after a period commencing in the 1970s and correlated with the widespread dismantling of industrial production in the Western world, many cities—even metropolises such as New York—experienced economic downturns and saw massive population declines.[10] Once this development took hold, Schmid argues, the underlying planning aspirations associated with the new metropolitan mainstream responded to the situation by attempting to stimulate growth, creating urban environments, even in former industrial areas, that were attractive to wealthier segments of the population. Such aspirations concern the desire to revitalize cities in light of the dwindling of the industrial culture that conditioned metropolitan culture, but to do so by relying on industrial production's urban forms. In part, this was done by repurposing old factory buildings into house offices or dwellings, albeit without the pollution, noise, or working-class lifestyle that used to accompany them.

Developing this argument further, we can identify a number of structures, including skyscrapers and other cultural and commercial flagship projects, as markers of the new metropolitan mainstream. These structures are metropolitan insofar as they borrow the paradigmatic forms of the European metropolises of the late nineteenth and early twentieth centuries—going back to the kind of urban environment of which the Eiffel Tower was part and to which Walter Benjamin's writings responded. They entail tall and dense urban architectural morphologies and infrastructures, but they also emphasize city life and are dotted with cafés, urban-cool artsy projects, and playgrounds. Despite their air of newness, these forms are employed without necessarily evoking a sense of pointing to the future or of being radical or provocative. Instead, they are broadly accepted and accommodating, an example being the idea of the "livable city."[11] They are what Schmid calls mainstream. Furthermore, these structures are historicist in their appropriation of urban forms that link up with past urban constellations, but they also promote dense urban planning paradigms that contrast with the postwar suburban sprawl. Moreover, we see different planning logics at play to push value markers such as real-estate prices.[12]

Attempts to steer urban processes in order to generate wealth are strategic instruments of more recent neoliberal urban policies, but they also continue long-term tendencies of urbanization that are characteristic of capitalist society more broadly. It would therefore be misleading to regard the word *new* in new metropolitan mainstream as a hard periodization denoting something that came after postmodernism. Such an understanding would buy into historicist notions of history as a progression of clearly defined ages, which is at odds with the significant longitudinal continuities involved in the phenomenon itself. At the same time, however, we argue that the new metropolitan mainstream points to central tenets of what is at stake in One World Trade Center and why the symbolic impetus of Libeskind's design was considered inadequate. Could it be that his design was simultaneously too gigantic and not gigantic enough in its efforts to speak to complex yet unifying symbolic registers in relation to the September 11 attack? As it was built, One World Trade Center conveys a more hands-on take on the relationship between the need for security and the possibilities for urban life and participatory engagement after September 11. These discussions reveal some apparent contradictions in current urban development that we see reflected in One World Trade Center.

In contrast to Libeskind's univocal symbolic gesture, One World Trade Center communicates well-known forms of gigantism from the nineteenth- and twentieth-century city, but it does so in a way that is generally appealing and unprovocative. The effects of this gigantism—which we argue is characterized by latency, understood here as a gigantism of the unobtrusive—become evident as we now turn to the digitized mediations that accompany the building, as well as the different ways in which it is embedded in digital culture. This concerns the tools used to realize the building's design, the films and media spectacles that accompany the visitor's experience of the building, and the way in which such visits, when uploaded to social media, feed into logics of datafication and networked analytics. These logics themselves operate on a gigantic scale and are part of the building's association with latent gigantism.

The Womb as Time Machine

Both the bomb that exploded in the Twin Towers' parking garage in 1993 and the plane attacks of 2001 have been interpreted as motivated in part by a desire to achieve the spectacular.[13] The fear of whether an attack would happen again has left its mark on the building's design and construction. This intense preoccupation with security can be read as an expression of hardline and even militant masculinity, which in the previous chapter we identify as characterizing the discourse around September 11. But a closer view reveals that its vocabulary of security is in fact a subtle engagement with the intricacies of the imagery of the container: the act of building on a gigantic scale at this site carries with it a stubborn sense of defiance, which has resulted in the building's resemblance to a gigantic bunker clad in glass. This is a form of material excess, a gigantism that literally thickens in the construction itself, not in a way that is hidden from view but in a way that is visible only if you look closely. From a distance, for example, the lower floors seem open and transparent because of vertical glass panels that stick out like fins. However, on closer inspection, these panels turn out to be decorative elements that hide from view a concrete construction so thick and seemingly impenetrable that it would withstand attacks from ground level by lorries or tanks.

The lower six levels of One World Trade Center also contribute to the building's securitization because they are mechanical floors and house no office space. Visitors and office workers enter through large doors from different sides and move up the building. The same aesthetic connotations apply to the vertical openings that penetrate the facade higher up. These openings are vents that can be used to let out poisonous air from inside the building in case of a chemical attack. The building thus fences itself off from the world and makes a virtue of its fear-driven aesthetics. Yet rather than being buried in the ground like most bunkers, One World Trade Center exhibits impenetrability and anonymity almost demonstratively. Turning its own container qualities into a defensive tool, the building seems to wait for an onslaught—anticipating and guarding against a future narrative of material inadequacy. As an imprint of the new metropolitan mainstream, the building's response to the possibility of an attack is not to hide but rather to protrude. It is not just visible: it is visually unavoidable—a securitized womb of gigantic proportions that, as becomes apparent when one enters the building, makes use of temporal conflation on multiple levels in the mediatized view that it offers. This paradoxicality—a blend of narcissistic grandiosity and Teflon-like mainstream anonymity—gives the building the oxymoronic quality of both gigantism and latency.

As with most of the commercial skyscrapers on Manhattan (apart from those named after the corporations residing in them), it is necessary to know which companies are based here in order to find them. Their location in this building is more or less undocumented from the outside, adding to the air of anonymity created by repetitious architectural forms and the hidden routines of office work. That office culture is mainly hidden from view, yet it indicates a sameness, even invisibility, that the grandiose building itself also exudes. The first tenant to move into One World Trade Center was the media company Condé Nast in 2014. Condé Nast attracts more than 164 million consumers with magazines such as *Vogue* and *Wired*, and it operates across a range of media platforms. The company occupies floors twenty to forty-four of the building's 104 floors. Although finance companies can also be found in One World Trade Center, the occupants currently on the rise in this formerly primarily financial district seem to be tech firms, high-end boutiques, and residences. A significant number of tenants of One World Trade Center are media, marketing, and entertainment companies. Six floors of the building, moreover, are occupied by the General Services Administration, an independent agency of the US government that includes US Customs and Border Protection. Floors 100 to 102 are occupied by the One World Observatory, which inscribes One World Trade Center in the tourist topography of the Ground Zero site.

The general public can enter the tower through its westernmost side entrance, turning away from the memorial site at Ground Zero (the building has no obvious front, and the people working in the offices here use different entrances). The public also can enter through the massively wide, skeletally white underground halls of the adjoining new subway station and shopping facility designed by Spanish architect Santiago Calatrava. In contrast to the notoriously phallic gigantism of Calatrava's landmark Turning Torso tower in the Swedish City of Malmö, noted in the prologue to this chapter, the subway station is a much more introverted space whose grand, white, rounded interior hall is marked by the upward and decorative movements of the white pillars and has religious undertones.

Inside One World Trade Center, the elevator, situated a couple of flights down into the base of the building, shoots visitors directly to the 100th floor, the location of the One World Observatory, an interior space equivalent to the outdoor viewing platform of the Twin Towers but closer to the "techno-monsters" Haraway argues are the offspring of visualizing technologies' coupling with the world. Although it is necessary to pay an entrance fee and pass through security checks, the elevator and observatory are linked to the underground public transport system that weaves beneath the street grid of Manhattan. The elevator and observatory are thus identifiable to some extent as public spaces (for those who can afford to pay). They therefore hint at an aspiration to embed the observatory in civic culture and allow forms of participation to take place, although they are surrounded by private office spaces that are accessible only from other, heavily guarded entry points to the building. Public and private intermingle here at the same time as they are also distinctly compartmentalized in a way that follows logically from the interplay between the desires for security and entertainment so evident in the building.

This coupling of security and entertainment is in tune with what cultural theorist Peter Weibel, in an essay published shortly after September 11, describes as the transformation of a panoptic principle into a pleasure principle.[14] In the face of forms of terrorism that operate to achieve maximum media coverage, visibility and transparency no longer connote security in the way they did in English philosopher Jeremy Bentham's notorious panopticon prison design from the late eighteenth century. This design was envisioned as a circular building that was lined by prison cells on the inner side and that had a watchtower in the middle from which the warden could observe the inmates at any given time. From this centrally located watchtower, the warden remained invisible to the prisoners, and as Michel Foucault famously extrapolated, the assumption was that prisoners would internalize this disciplinary gaze because they would never know when they were actually being watched.[15] Today, Weibel's diagnosis of the alliances between surveillance and security seems even more pertinent, not least in social media's participatory digital publics ripe for datafication and calculation.

The architecture of One World Trade Center is indeed characterized by an intermingling between the idea of the bunker, which fences off the building's interior, and an engaging, softer vocabulary that encourages participation by inviting visitors to interact with their surroundings through their smartphones. We can thus see this participatory mode as indicative of a much more encompassing and indeed gigantic surveillance regime, which involves more or less calmly embedded forms of tracking and data collection. The security measures at the One World Trade Center building echo this complexity by enforcing a regime of transparency as security. The building itself is cloaked in anonymity, and the observatory's audience is guided through safety checks (including metal detectors) as people enter.

At the same time, the viewing platform caters to the exhibitionist and voyeuristic pleasures of social media culture, allowing people to check into the One World Observatory virtually and to post images from the site and the views that it offers. The tower can in this way be seen as a transmitter, less on account of the 408-foot antenna at its top and more because it is a hub in the tourist economy that constantly leaks information and metadata by and about the people in it.

Before we rush to the top, let us remain inside the elevator for a moment. The elevator ride functions as a prolonged entry into the observatory. It offers different mediatized forms of experience, many of which cater to visitors looking for motifs to upload to their social media profiles. These mediatized experiences are examples of the interplay between an internalized regime of security and a form of publicness that is digitally enhanced, and it unfolds inside the publicly accessible part of the building. This interplay also interlinks with the building as part of an overall vision of urban life that is trapped in contradictory forms of temporality. The ride up is accompanied by a film that is projected onto the walls of the elevator and tells the story of Manhattan's architectural development, from the first colonial settlements onward. The storyline of the film collapses time and space by means of a fast-tracked linear temporal engagement with the Manhattan context as visitors move upward in a similarly teleological movement to reach the observatory. It creates a sense of standing on the site of One World Trade Center itself and an experience of the city's over five hundred–year history from the point of view of this site. In this grander storyline of the interplay between erection and destruction, the elimination of the Twin Towers is somehow naturalized within a greater narrative of architectural prowess and progression. Indeed, the fast-forward temporality of the film collapses five hundred years of building activity from swamp to village to a skyscraper-clad skyline. For just a couple of seconds in the film, which span the period between 1972 and 2001, we watch the South Tower of the original Twin Towers. Then it disappears, and we see the scaffolding of the current building, until we reach a time that overlaps with our present and can step out into the observatory to engage with the current view.

The film is intriguing in the way it articulates a vertical movement through space that is simultaneously a linear timeline and a viewpoint that moves. The film elevates a narrative of progression of Manhattan's architectural culture as a literally upward movement, where—despite the (seemingly

One World Trade Center elevator. Photo by the author, 2018.

necessary) destruction of some older buildings to make space for the city's general upward progression—the skyline becomes taller, grander, and more beautiful as time progresses. As was pointed out when the elevator film was released, however, it does not give a historically accurate rendition.

Presumably, this was done to include as many significant landmarks as possible, catering to the tourist gaze of which the tower is part.[16] In this elongated time span, moreover, September 11 occupies only a few seconds, and the event's catastrophic connotations are smoothed out by being written into a longer story of construction and destruction, signifying the architectural change and urban rejuvenation that is characteristic of New York City.

This is a very different statement than was involved when Libeskind proposed to make the height of the building 1,776 feet to mark the year of independence in a grand symbolic gesture, as if he were arresting not only time but also space. The elevator ride and the observatory can be used as a highly tangible vehicle for understanding that the gigantism of One World Trade Center should be seen as both more and less gigantic than the building proposed by Libeskind. It recasts the narrative about Ground Zero and the destruction of the World Trade Center in a more optimistic way as part of inevitable urban processes of construction following destruction that appear to follow a naturalized linear path of upward movement. Whereas Libeskind's original design was haunted by the site's recent history, One World Trade Center and its observatory are curiously devoid of any mention of it. The tower engages with the city's history in a broader historical frame, and its scope is the long history of the site (although it goes back only five hundred years, and as such presents an anthropocentric and colonial history from a distinctly Western point of view). As public media reported when the observatory opened: "Here in the observatory at the top of One World Trade Center, we're looking toward the future."[17]

The elevator ride can thus be seen as a way of literally rising above the immediate past to reach a viewpoint that is unrestricted. In the previous chapter, we engage with the complexity of such a God's-eye perspective onto the urban fabric from the viewing platform of the Twin Towers. Here, what concerns us is the way in which temporality is dealt with in a proclamation of prophetic abilities. It is tempting to contrast the upward stretch of the elevator ride against the museum and memorial pools. The latter are concave structures, containers of the past, and the tower points toward the sky and the future, an emblem of the linearly progressing, measured time of modernity. Yet the elevator ride, as we argue, also points to the fact that a different conception of history and temporality is at work in the observatory and memorial spaces of the Ground Zero site more generally—one that expands time.

In the elevator, the technique of lining the walls with monitors that look like glass windows removes any sense of orientation and contributes to the overall feeling of a broad mediatized present, in a way familiar from casinos or similar modern entertainment institutions that build worlds within worlds where time is designed to evaporate.[18] The elevator inside the skyscraper as a womb with a view offers this view in a mediatized conflation of temporality while playing with transparency as a formal gesture in a way similar to many skyscrapers and in dialogue with the likewise faux transparency of the building's facade. From the inside as much as from the outside, the building thus mimics transparency and yet returns in both instances to confinement. Moving rapidly upward through the building at a speed of approximately thirty-seven kilometers per hour, while simultaneously fast-forwarding in time through more than five hundred years of the history of Manhattan, implies a conception of history that can take spatial form.

This resonates with media theorist and architect Mario Carpo's description of a bakery in his childhood town in Italy. Carpo tells the story of a bakery named Il Forno Moderno (The Modern Bakery), which closed and reopened in the 1980s as L'Antico Forno (The Old Bakery), and shut down in the twenty-first century. Then the bakery building was turned into a cell phone shop named Global Roaming (written in English). He connects this narrative with questions pertaining to time, history, and the use of digital technologies in architecture and culture more generally:

> The digital revolution that marked architecture at the end of the twentieth century may have been the first self-proclaimed revolution in recent Western history to take place for the most part without, and outside of, any established philosophy of history (to revert to my bakery analogy: no future, no past—just "global roaming"). This may seem a contradiction in terms, as the notion of a revolution implies that something is being disrupted; and the primary object of a revolutionary disruption, in the Hegelian tradition, used to be a historical process. But a revolution is a modern ideological construct. If you try to make one in a postmodern environment, odd things may happen—as they did in this case.[19]

Carpo here articulates what we address in this book as a set of temporal categories that link up the modern (as in the modern bakery, with connotations of linearity, progress, and progression), the postmodern (as in the old bakery's play with the word *antico* during a period that witnessed a literal renaissance of historical forms), and something that is simultaneously latently present and gigantic in its ubiquity (as in global roaming).

Unlike traditional linear histories' emphasis on progress and change as driving history forward through dialectics and potentially toward revolution, the condition of global roaming describes a broad present that conflates past, present, and future while simultaneously containing linear conceptions of history within it. The broad present thus allows the idea of linear history to exist in parallel with other temporal categories, and it therefore also does not represent an end of history in a posthistorical sense. The temporal category Carpo invokes as global roaming is thus more akin to the latent gigantism that we see in One World Trade Center than to the postmodern epistemologies and metaphors of networks and simulacra we discuss in the previous chapter. In contrast, latent gigantism speaks to a sense of simultaneity, temporal overlay, and horizontal spatial distribution—for example, facilitated by cell phones and by the digital infrastructures that the observatory caters for and encourages its visitors to engage with. Significantly, the elevator film can be found on video platforms such as YouTube for repeated viewing, some of which were clearly made by tourists capturing their ascent on their cell phones. This opportunity for displaced and off-site viewing adds yet another temporal layer of repetition and broadening of the present moment, and the number of such uploads testifies to the popularity of this experiential mode and the extensiveness of the number of people who by chronicling their experience of the tower inscribe themselves in algorithmically calculated publics.

If the Eiffel Tower was our Modern Bakery, epitomizing the pinnacle of industrial prowess in the 1880s, then the Twin Towers' neo-Gothic doubling is our Old Bakery, and One World Trade Center's exhortation to See Forever is this book's equivalent of global roaming. When it comes to the mediatized logics of the observatory as well as the antenna atop the building, however, the One World Trade Center tower not only roams but also leaks information. Let us therefore explore how the mediatized tourist experience of the building develops as we exit the elevator and get closer to the actual observatory space.

A Leaking Container

The integration of mediated fictional views of the city into the experience of the elevator ride makes the tower participate in a contemporary tourist culture of augmented entertainment. The focus on time and its passage in the mediatized tourist spectacle of the One World Observatory continues as we exit the elevator and enter the observatory space on a gray morning in May. Together with the rest of the audience, we stand in front of a thick white wall with a three-dimensional pattern that looks like the tower buildings that are ubiquitous in Manhattan. This wall is used as a projection screen for another filmic mediation of time, architecture, and the metropolitan materiality of life in Manhattan. The short sequence takes the form of a montage suggesting a day in the life of New York City, reminding the tourists in the tower of typical figures of the life and fabric of this city, including famous tourist topoi such as Times Square, yellow cabs, the river, iconic skyscrapers, and typical residential forms of architecture.

With an image quality that does not necessarily rise above what most tourists are able to capture with their smartphones, the film starts with a view down onto street level and pans up toward the tips of the skyscrapers towering above. It then zooms in on various typical aspects of city life, suggesting character traits specific to New York—the bridges, historic buildings, or more general cultural figures, such as the diversity of the people and the street life of the city's different boroughs. Another theme is the subway and its colorful signage. These visualizations all point to stereotypical metropolitan forms of which New York City is a significant originator, as if the film had been made to market the building (and indeed the city of New York) as a vital hub for the new metropolitan mainstream. Yet the film also presents an example of how such filmic montage techniques can today be composed and widely distributed online by anyone with a smartphone, which gestures toward a highly distributed and participatory mode of experiencing.

As a filmic montage that pastes together typical historical metropolitan motifs in an upbeat story about the unique, dense, busy, yet accommodating and energizing New York lifestyle, the film appropriates a vocabulary of the historical avant-garde. The montage technique makes the film, in relation to both form and content, a kind of filmic version of the new metropolitan mainstream, which we discuss earlier in relation to architecture and urban planning as a preoccupation with reshaping typical forms of metropolitan culture into a more beautiful, smoother, healthier, more livable, but arguably also flatter new version. The layering of the view makes that view denser by means of temporal suspense, spatial multiplication, and processes of association, creating a broad present that revolts against punctuality and precision and invokes an almost dreamlike state—not dissimilar to the dream of the future Copenhagen cityscape described in the prologue, in which past, present, and an imagined future collapse into one.

The grand finale comes when the wall/screen on which the film is projected turns out to be mobile. It slowly rolls up to reveal a glass window and the actual view of Manhattan from the top of the tower. Even though the tower was clouded in fog (the fuzzy view is emphasized by Maria Finn's drawing at the start of this chapter) and we saw nothing but a thick white cloud and large raindrops clinging to the outside of the window, the effect was dramatic. However, the audience is not granted this view for long. The screen rolls down again while a voice invites us into the observatory itself, accompanied by the injunction to See Forever as a slogan for the position granted by the viewing platform. In this way, it suggests a collapse of time and space, while at the same time the insistence on the eternity of human vision indicates a broadening of the kind of viewing that takes place in the tower. At a point in time when ecological and political crises have rendered optimistic sentiments about the future cruel, this aggressive sense of futurity introduces a sense of hope that the current building is capable of eternal life, in contrast to the Twin Towers it has replaced.

Such assumptions are characteristic of the technological optimism of modernity's historical and linear time of progression, but at the same time

See Forever Theater film. Photo by the author, 2018.

it here evokes the sensation of a broadening present. The sense of spatial and temporal overlay and distribution is again emphasized by the many YouTube uploads featuring this film. In these uploaded videos, which all capture the same experience with small differences in viewpoint, we can see other audience members recording the film on their cell phones too. Repetition with a difference, coupled with participatory engagement, here forms an experiential mode that is both flattened and widened out by being posted on the same platform next to thousands of similar experiences. Montage, which we saw in Benjamin's reading of the Eiffel Tower as emblematic of the mechanical age of the engineer, has here become an act of mediation in which seemingly everyone who has entered the observatory can participate, adding personal versions within the smooth framework provided.

Indeed, in ways to which networked media are prone, this participation is prompted, and it comes with a certain sense of the compulsory, as we discuss in chapter 1 in relation to the social media response to the 2015 terror attacks in Paris. At the same time as literal, so-called Instagram museums are currently emerging, offering backdrops for photos and films that are suitable for social media, more traditional museums and entertainment venues are also finding themselves responding to and engaging with these designed modes of digital consumption. In this way, they transform the traditional museum's insistence on material authenticity and preservational logics into an engagement with other forms of authenticity that pertain more to visual likeness and emotional or atmospheric sensations.

Now as we finally enter the actual observatory space, which promises a 360-degree view of the surroundings, the mediated trajectory of the tourist gaze continues. A series of ramps guides the visitor past more opportunities to engage with the space through different forms of participatory media installations. These constitute a trajectory that leads to the 360-degree window, through which the physical topography of Manhattan can be explored at the visitor's leisure. When we first visited, one option was an iPad with a movable picture of Manhattan (identical to the view through the windows of the observatory) that gave explanatory descriptions of the buildings, with possibilities to zoom in and out of particular details of the cityscape. Another option was to be photographed in front of a picture screen depicting the view blocked by that very screen.

These different forms of engagement with the view all speak to a flattening of visual experience and representation that comes from the uniformity

Layers of media. Photo by the author, 2018.

of the platforms through which the views are experienced. Although visitors are prompted to engage and interact with the view through these various digital renditions and interfaces—which in this way solicit individual responses—the form of participation that is encouraged will inevitably render the forms of response more or less identical.

This nudging of the visitor with respect not only to *what* kind of view of the city is possible but also to *how* it can be captured and disseminated speaks to a different enactment of power than the one Certeau identified from the viewing platform of the World Trade Center's North Tower. Although no less mediated, that viewing platform provided a view that was mediated primarily by the sheer height of the building and whose power resided in the vertical gesture of rising above the streets of the city.

In contrast, the One World Observatory also offers a horizontal experience in which time (the linking of the experience to the site's past) and space (the different filters that are offered through which to *see* the urban space of Manhattan, such as glass and screen mediations) merge in a way that uniformizes and directs the individualized experience—at the same time as there is a gesture toward a larger sharing or shared experience at work, as we discuss in the next chapter. For instance, the cell phone in our hand comes to form a sensory extension that allows us to photograph or film the experience and send it to loved ones across the globe, in this way making them present in the tower with us or enabling us to watch the recordings again later. Through these media effects, which collapse time and space, the viewer's understanding of the *here* that the observatory offers refers to a *here* that reaches far beyond the viewing platform, both temporally and spatially, by encompassing the wider network of social media through which the views are disseminated. In this way, a particular kind of gigantism is constructed that is characterized by latency, both spatially (embodied by the reach of the network) and temporally (reflected in the invitation to See Forever).

What the mediatized spectacle of One World Trade Center allows is a situation where nothing—not even varying weather conditions or the range and capability of the human eye—constitutes a limitation that the mediated experience of the view cannot circumvent. This insistence on freedom from embodying conditions constitutes a form of gigantism of its own. As a container, the tower is continuously transmitting; indeed, it invites this informational leak through the way it engages its visitors. The invitation to participate and thereby leak information is in stark contrast to the outside appearance of the building as an impenetrable bunker of gigantic proportions. It speaks to the complexity of the architectural vocabulary at work, of which its mediated design is a key point.

New Visual Orders

In Daniel Libeskind's original drawings of his proposed building for Ground Zero, the spire located on one side of the building strongly resembled the arm of the Statue of Liberty. The antenna atop what in this early version was called the Freedom Tower was thus heavily imbued with the symbolism of freedom and the frontier (whether the Wild West, space, or cyberspace) by making the wireless signals emanating from the structure a natural (if invisible) continuation of Liberty's outstretched arm. The Freedom Tower would have symbolically joined the strikingly visible (the monumental tower) with the invisible (wireless communication), thereby also allowing an interpretation of the antenna as an extension of the reach of US ideologies.

Libeskind's designs—which offer a vocabulary in which invisible cultural meaning and affect are brought into the light for all to see—have been criticized for having too crude or contrived a symbolic language.[20] His design suggests that what is intangibly felt can take physical form as a

Looking out of the One World Observatory window. Photo by the author, 2018.

The One World Observatory

The World Trade Center site in relation to the Statue of Liberty. © Studio Daniel Libeskind.

bodily experience in a gigantic format on and around Ground Zero. However, during the journey from the architectural drawing board in the initial design competition in 2003 to the built structure that went up a decade later, something happened that indicated more complicated cultural issues were at work than the search for a timely architectural style or design approach to the current design problem. There are thus more latent visual orders at play in this site that did not find form in Libeskind's symbolic design but that are present in One World Trade Center. Perhaps this is because the design as it was actualized evades questions of architectural form-finding when compared with Libeskind's proposal. The mainstream anonymity of the World Trade Center and the mediated flattening of the way it is experienced in the observatory thereby embody latent rather than semantic gigantism.

The architect who eventually replaced Daniel Libeskind as the lead designer on One World Trade Center, David Childs, represents the large American firm Skidmore, Owings, and Merrill. The firm is known for designing sleek commercial buildings, such as glass-box skyscrapers, rather than being associated with the expressive personal style of starchitects such as Libeskind. The changed design of the building's spire illustrates this

Statue of Liberty with One World Trade Center. © Steve, 2014. Wikimedia Commons.

difference and move from an emphasis on semantic to latent gigantism. Instead of Libeskind's swirling reflection of Liberty's outstretched arm, the revised spire is in dialogue with more traditional New York architecture, such as the Empire State Building, the Chrysler Building, and indeed the antenna at the top of the original World Trade Center North Tower. It thus offers a subtler historical reference and a more explicit function than Libeskind's design, which was motivated by symbolic intent.

This does not imply that the antenna's design on One World Trade Center does not carry cultural meanings. Indeed, the antenna arguably summarizes the different forms of gigantism at work, making One World Trade Center not only recognizably linear and phallic but also at the same time a very literal and highly securitized gigantic womb with a view that invites its visitors to engage and interact in highly controlled ways. The antenna is thus an example of how, when One World Trade Center opened in 2014 as the finished result of a decade of negotiations over what kind of structure should replace the Twin Towers, the building had stakes in much wider cultural negotiations about how to make visible—how to name, erect, create, and even show—questions that are particularly loaded for this specific urban site.

The One World Observatory

Tall spire. One World Trade Center, including its spire, measures 1,776 feet (541 meters), a deliberate reference to the year when the United States Declaration of Independence was signed. © IIP Photo Archive / US Department of State/ Flickr.

One World Trade Center as it now stands thus emerged out of lengthy negotiations as one building to contain them all—including the interests and emotions of the previous buildings' owner, the current and potential occupants on Lower Manhattan, a nation preoccupied with the impact of the terrorist attacks of September 11, and the global interests of a world whose eyes rested on New York and on the undertaking to rebuild on this particular traumatic site. The building's exterior form appears as a neutral compromise whose sharp edges have been rounded through consensus. Through precisely this compromise, it conveys past configurations in a visible paradox of grand proportions that is a continuation of both the overt gigantism with which the site is associated and the more latently present forms of gigantism that we see gaining a foothold in post-9/11 culture.

Alongside One World Trade Center's visible manifestation of a skyscraper aesthetics—seen, for instance, in the way the tower boldly inscribes itself in the height contest of what can be measured—the tower also alludes to other visual modes that are at play much more calmly, yet no less influentially, in the building. They connect with temporal regimes that are horizontally distributed (for example, through digital information technologies and media practices) and that are characterized by temporal overlay and synchronicity. In this way, One World Trade Center as architecture and media points toward more fundamental questions of visual order in the early twenty-first century that we argue connect with latent gigantism. Let us now consider the extent to which this latent gigantism applies both to the mediated experience the visitor gets in the tourist topography atop the tower and also to the building's design.

Computational regimes of form-finding are at work in contemporary architecture in the wires that cut through these buildings, in the wireless signals that move invisibly between floors, and in ways that architecture is imagined and designed. Indeed, the architectural design process today is increasingly influenced by the properties of the digital tools used to imagine and design buildings. There are many examples of this effect on architectural form, sometimes more noticeable or self-consciously employed than others. In New York City, the complexly twisted and curvy building from 2009 of the Irwin S. Chanin School of Architecture at the Cooper Union for the Advancement of Science and Art is a good example.[21]

The influence of digital architectural tools on architecture speaks to questions about how a structure such as One World Trade Center reflects the conditions of production at the time of its construction. This recalls our discussion in chapter 1, where we consider Walter Benjamin's interpretation of the history of iron construction and its use in architecture in the late nineteenth century. Iron construction was key to the principles that defined the Eiffel Tower's architecture as new, avant-garde, and even shocking. Yet at One World Trade Center, the algorithms have not been asked to bend material beyond the apparently physically possible in 3D renditions, as has been the case in more expressive and iconic starchitect architecture. Rather, they have been streamlined for optimal functionality, cost-effectiveness, and consensus.

However, the embedment of the building's design in digital culture is no less pervasive. Like the Eiffel Tower, which was described as simultaneously a gigantic structure and a knickknack, One World Trade Center, understood as metropolitan mainstream, can be seen as an expression of a design aesthetic that has much in common with many of the smaller-scale digital technological devices we hold in our hands. The formal expression of many items—such as smartphones, voice-controlled smart speakers, and other devices currently permeating households—likewise seems rounded by design compromises and optimizations. What is at stake here, we argue, is a catering to mass consumption, which has led to a flattening of visual display and come with a uniform aesthetic that is smooth and pleasant to touch and look at. At the same time as the life cycles of such gadgets are increasingly short, a nostalgia for older technology that exudes a sense of durability can also be seen—for instance, in Apple's indebtedness to German architect Dieter Rams's designs for Braun in the 1960s and 1970s, which are visible in a range of products, including the now defunct iPod, the MacBook, the iPad, and the iPhone.[22]

The example of Apple's design inspiration is notable not only because of the currently seemingly ubiquitous presence of Apple products, particularly in affluent parts of the world such as New York City, but also because in the present context of our reading of One World Trade Center architecture, there is a specific connection to Skidmore, Owing, and Merrill as epitomizing the skyscraper era in the early twentieth century. Rams's designs grew out of the Ulm School of Design as a postwar successor to the Bauhaus, as well as his background working for the architectural company Apel, which at the time was working with Skidmore, Owing, and Merrill. This company connection brought Rams into contact with the works of Mies van der Rohe and Walter Gropius in Chicago and New York, a contact that is said to have had a lasting influence on Dieter Rams's work.[23]

Similarly to the One World Trade Center tower, these technology designs harbor the duality of a desire for emancipation—to reach beyond themselves as connected and leaking physical objects (such as the cell phone or smartphone's promise of global availability, represented by the antenna that works almost like the antenna of an insect that extends it sensory capabilities)—and an embeddedness in the past in terms of their design heritage. In fact, we may say that these gadgets embody the new metropolitan mainstream no less than One World Trade Center and give shape to the same latent gigantism. It is difficult to overlook the resemblance in the aesthetic appearance of these objects, which is calculated to facilitate the best user experience with minimum effort, while their inner workings remain black-boxed and (unless hacked) are susceptible only to minor adjustments by the user. Perhaps it is therefore not surprising that there seems as yet to be no great cultural theorist—like Benjamin, Barthes, or Baudrillard—who has built an expansive theory around this form of architecture.

At first glance, we might question whether we have yet seen algorithms and protocols influence physical design in the way that iron construction influenced the design of the Eiffel Tower. Yet we argue that we do in fact see this influence on a large scale in contemporary buildings such as One World Trade Center. This is not only evident in how One World Trade Center's design is tied in with the highly digitized production of architecture in the present day or how media technology has been designed to be experienced in the One World Observatory, as we have explored in this chapter, but also in how it cultivates an aesthetic as smooth as an iPhone—as if the question of scale were irrelevant to designers who work with the gigantic, whether through monumental size or global range. Mario Carpo and others have shown that this evidences the way digital technology lies at the heart of the forms that contemporary architecture takes.[24] We have aimed here to trace these implications in reverse.

Temporal and Material Excess

Reading One World Trade Center's architecture as a streamlining toward the new metropolitan mainstream in gigantic format, in relation to both vertical and horizontal gigantism, points to wider discursive slippages in cultural theory concerning fundamental categories of history, temporality, and materiality. We may now ask how the forms of gigantism made manifest in One World Trade Center tie into twentieth-century discussions about history (and its end), which we broach in the previous chapter. The question of what made the One World Trade Center design appropriate for the Ground Zero site is an epistemological problem related to conflicting understandings of history and time. Understanding this problem may help us to understand why Libeskind's design was ultimately deemed less appropriate for this site. Latent gigantism is linked to an understanding of temporality where extreme manifestations of emancipation from the conditions of embodiment (as embodiment reaches beyond itself through gigantic structures or communicative signals) are brought together with an extreme reliance on past forms but without making these relationships directly visible or charging them with deeper symbolic meaning. Let us therefore consider the extent to which this excess (which leads to both temporal and ontological flattening) also has parallels in theoretical writing by returning to the previous chapter's questions of postmodernism, posthistory, and the sense of a future posterity or broadening present.

The time when the Twin Towers were erected in the 1970s is concurrent with the onset of architectural postmodernism, but it also corresponds to the onset of postmodern currents in philosophy. Well known in this context is French philosopher Jean-François Lyotard's identification of the fragmentation of grand narratives—including the narrative of history as progress—in his work *The Postmodern Condition: A Report on Knowledge*.[25] The prefix *post* (meaning "after") gives a sense of something that comes after modernism while still being attached to it. At the time, poststructuralists, Marxists, and neoconservatives alike engaged in different ways with the sense of reaching an end of something, perhaps most notably and polemically captured in American historian Francis Fukuyama's declaration of the end of history in 1992.[26]

As we discuss earlier, the terrorist attack on September 11 was seen by many commentators as a wake-up call to a cruder reality—a parting with notions of cultural relativity in a way that was already embedded in the complex thinking of figures such as Michel de Certeau and Donna Haraway. At the same time, however, 9/11 became linked to a turn to realpolitik in tune with the posthistorical interpretation of the ending of the Cold War in 1989 as the end of a conflict to end all endings. The destruction of the Twin Towers and the bursting of the dot-com bubble both took place in 2001, and they could be said to indicate an ending to the paradigm of the end of history. The sense of destruction in the attack and in the War on Terror that followed also furthers dystopian cultural articulations that are concerned with an end in material terms—not just through war, conflict, and division but even as the elimination of humankind and the globe. Cultural theorist Mikkel Bolt Rasmussen has recently polemically commented on the commonplaceness of the notion of the end:

> "The end of history" obviously carries more weight today. *This* is the end we have embraced and in which most of us are spending our daily lives (in the global West, at least). It comes equipped with lattes, iPhones, and a pair of Vetements jeans. This is the *lingua franca* of both politics and academia, either in the form of some kind of melancholic lament for the past or just the consensual administrative talk of politicians and bureaucrats alike, "where there is what there is." Simulacrum all over again. We are trapped in representation. . . . We seem to be living in a zombie afterlife of modernity in which its forms and institutions still exist but have been completely hollowed out.[27]

This dismal analysis, which bears testimony to how a hope vested in revolution (as a Hegelian move forward) has capsized into a zombie afterlife, resonates with the temporality that we address in this book as latent gigantism. The underlying temporality is one that has lost faith in linearity and in the radical break traditionally connoted by the notion of revolution (such as in the decisive crisis or catastrophe, as we discuss earlier).

Significantly, we find such latency in fundamentally different strands of contemporary scholarship that engage with different subject positions. Saidiya Hartman, for instance, has recently gathered histories of a group of young black women in larger cities in the United States in the early twentieth century where she recognizes what she calls a "revolution in a minor key" that unfolded in the city with these young women as its agents.[28] This conception of the temporality of revolution—not as a break but as a prolonged moment that takes place every day and that can be dug out of the archives of the past and identified in hindsight—also bears the mark of latency, albeit in a very different way. What Hartman identifies as a "revolution in a minor key" occurred as a "consequence of economic exclusion, material deprivation, racial enclosure, and social dispossession; yet it, too, was fueled by the vision of a future world and what might be."[29] Bolt Rasmussen's analysis illustrates the condition of latency and ennui from the

Western perspective of the privileged who can afford an iPhone and a pair of designer jeans and for whom the revolution has become an abstraction. The latent temporality that Hartman's methodology identifies speaks from the position of the dispossessed and excluded who cannot afford *not* to hope for a better future. Yet both can be read as engaging in different ways with a latent temporality that struggles to come to terms with nostalgia or hope for a better future—whether through revolution as a great overthrowing or through the everyday resistance that slowly pushes boundaries.

In architectural terms, our analysis of the One World Observatory has identified a temporal flattening mirrored in the mediatized experience that the observatory offers, whereby time is conflated into a broad present. However, rather than seeing this condition as a simulacrum—an entrapment in representation, as Bolt Rasmussen does—we see it as a different kind of entrapment in which these temporal flattenings are paralleled by material latency. Latency in gigantism has to do not only with temporality and a sense of a broadening present, perpetual crises, and everyday revolutions; latency also has a material quality. As we now discuss, this material quality impacts what we can learn about latent gigantism from theories that emphasize the relationships between humans, technology, and nature, for example, given in the work of the French philosopher Bruno Latour (1947–), whose work we turn to in more detail in the next chapter. What we are interested in here is the degree to which they risk building relational utopias of their own.

An illuminating discussion can be found in the work of Indigenous feminist writer Zoe Todd. She declares Latour her intellectual hero but is critical of his notion of Gaia, a holistic conceptualization of the planet-thing-living-being-and-human as one intricate relationship and of the earth as a symbiotic system where all living things co-evolve with their surrounding geological and atmospheric conditions.[30] Todd criticizes the underlying understanding of newness in the notion of climate as a common organizing force: "What struck me here was the unintentional (even ironic) evocation of theories about the climate as a form of *aer nullius* which it often becomes in Euro-Western academic discourses: where the climate acts as a blank commons to be populated by very Euro-Western theories of resilience, the Anthropocene, Actor Network Theory and other ideas that dominate the anthropological and climate change arenas of the moment."[31] Significantly, Todd appropriates this *aer nullius* (*nullius* meaning "belonging to no one") from Indigenous studies scholar Glen Coulthard's notion of *urbs nullius* or *terra nullius*,[32] a critique of the way scholars and urban planners in North America address urban spaces or land as commons that can potentially be populated by everyone, thus disregarding Indigenous people's rights. To our mind, such notions of assumed "blank commons" represent a gigantism that performs a flattening which potentially glosses over other calls for human and ecological justice, and we need to carefully consider what happens when gigantic inequalities are masked as universalities. We moreover must be careful not to bypass the important postcolonial and feminist scholarship that pays attention to how claims of universality and "commonness" may gloss over differences and inequalities and in fact reveal themselves as a latent gigantism.[33]

Media theorist Joanna Zylinska's *The End of Man: A Feminist Counterapocalypse* makes such an argument, when she suggests that the Anthropocene narrative evokes imagery of the apocalypse. Often, this narrative signals the expiration of Western power structures and what she calls "White Christian Man," but the anthropocentric orientation of this narrative also risks reinstating universal man as a figure implored to act on the basis of scientific objectivity and in the face of imminent crisis. In her aim to carve out a position for ethical orientation and participation in this situation, Zylinska reminds us:

> Indeed, there is no way of unthinking ourselves out of our human standpoint, no matter how much kinship or entanglement with "others" we identify. It is also next to impossible to abandon our human mode of perception and suspend the material and epistemological subject-object divisions we humans introduce into the flow of matter (including our primary positing of what surrounds us and makes us as "matter")—notwithstanding the misguided even if well-intentioned attempts to think like a bat, walk like a sheep, or float like a jellyfish. Rather than fantasize about some kind of ontological "species switch," the ethical task for us humans is not only to see ourselves as contaminated but also to account for the incisions in the ecologies of life we make, for the differentiations and cuts we introduce and sustain, and for the values we give to the entities we have carved out of these ecologies with our perceptual and cognitive apparatus.[34]

Indeed, any discourse of flatness is itself oriented.[35] This means that if we take terms such as *neutrality*, *commonness*, or *hybridity* for granted, we ignore the fact that architecture as much as technology is part of a larger discursive formation of differences, relations, and dependencies, with particular affordances that are, for example, gendered and racialized and that most often privilege the already privileged, loquacious, or able-bodied. We need to be wary about how these affordances constitute the concrete urban spaces where the cultural life of our cities plays out, as well as the possibilities for ethical orientation and participation they enable. This is exactly what Saidiya Hartman succeeds at in *Wayward Lives, Beautiful Experiments* by focusing on specific lives and their interrelations, while also showing the difficulties of escaping the temporalities of gigantism implicated in any such endeavor. This is also a challenge we face in this book but we address it on a different scale and without the sophisticated empirical work of a scholar such as Hartman. We focus at the level where the architectures and sites are construed as belonging to everyone and no one. The differentiations we outline imply the intersections of the workings of architecture and digital culture more than the actual lives that are played out with and within the buildings. We are concerned with the structure of how architecture, technology, and human life contribute to forming injustices, inequalities, and unsustainabilities but also prevails in concrete spaces of connectivity and connection in cities today. We will unfold this challenge further in chapter 4 through a closer consideration of the ideas of commonality that develop in light of latent gigantism.

Maria Finn, *Unfinished #21*, pencil on paper, 29 × 42 cm, 2018. © Maria Finn.

4 The Ground Zero Site: Calmly Common

Prologue: Trump Tower

New York City, November 2018

I look at pictures on the internet of Donald Trump's penthouse condominium on the sixty-somethingth floor of Trump Tower in New York City. They reveal a nouveau-riche setting with quasi-imperial references to the grandeur of the court at Versailles under Louis XVI. But we are clearly not in seventeenth-century France, and the gilded furniture and marble-and-mirror décor are punctuated by signs of everyday life in the early twenty-first century: a telephone, a box of Kleenex, an open MacBook. The thick white carpet is spotless; it looks as if no one has ever walked on it. It makes me wonder: Who does the cleaning? Who has the direct telephone number? Who has the runny nose? The apartment is inaccessible to the public, but the atrium at the bottom of the building is in principle open to anyone. Trump agreed to include a public space on the lower floors of the tower in exchange for being allowed to build a taller tower than the building regulations prescribed. We may say he traded height for public availability back when the Trump Tower came into being in the early 1980s.

I visit Trump Tower on a cold and dark afternoon in November 2018 to investigate what kind of publicness Trump's tower on the high-end shopping street of Fifth Avenue might in fact entail. The gilded front entrance is guarded by New York City police officers, some holding semiautomatic machine guns. After going through security checks similar to those encountered at airports or One World Trade Center, I step into the confusing interior of a four-story-high atrium clad in shiny pink marble, tinted glass, and gilded metal. It makes me think of Ada Huxtable, the famous *New York Times* architecture critic, who characterized the atrium of the Trump Tower as a pink marble maelstrom. Yet the reflective, meat-colored surfaces give the space a surprisingly warm, womby, and comforting feeling, a bit like the calorie- and sugar-rich comfort food served in the restaurants here. The warm light of the atrium contrasts with the image of the black, unwelcoming tower I had seen from the outside. The contours of the tower were difficult to discern against the dark November sky. Apart from a couple of lit windows, from the outside the tower looked completely unoccupied, giving away as little as the quietly dispersed people in the atrium, none of whom disclose their reasons for entering Trump Tower that day.

The gold and glitz and the sound of cascading water in an interior waterfall give the space a certain pomposity, although the architecture looks worn. The low ceilings, displaying a pattern of square plasterboard, hint at rationalized and cheap prefab building techniques behind glossy surfaces. The persistence of the word *Trump* in the atrium space—Trump Store, Trump Café, Trump Grill, Trump Boutique (selling more upscale Trump merchandise than the knickknacks in the Trump Store), Trump Bar (note that Donald Trump is a teetotaler)—makes no attempt to shield the branding of the space. The only pictures of the atrium I have seen before were from August 2017, when Donald Trump, bathed in the sharp light of a projector, gave a press briefing here. The briefing happened just after a car attack in Charlottesville, Virginia, where a young woman was killed by a neo-Nazi in an assault on a crowd of demonstrators. Donald Trump's remarks at the press briefing in Trump Tower notoriously emphasized a relativistic view of what had taken place. I am not sure whether the heavily securitized entrance to Trump Tower makes me feel safer or less safe than anywhere else on Manhattan.

The knickknacks sold on the lower level of the building, next to the passageway leading to the public restrooms, include Christmas tree decorations with pictures of the president. The Starbucks coffee shop on the second level features a picture of Ivanka Trump kissing one of the brand's paper cups. The photo is signed and has a handwritten note stating that this is Ivanka's favorite place to drink Starbucks coffee. The sexualized connotations of this picture aside, all the shops ooze an intimacy that the formality of the pictures from the inaccessible apartment defy. I have to acknowledge that, having entered the space, I unwillingly consume the Trump brand, but something else is at play, too, that I had not expected and that I tell myself merits my brief presence in this space. It is a marked conflation between public and private—between the aspirations of grandeur and gigantism and the latent, creepy presence of junk and objectophilia in the open (if highly controlled) atrium space of the tower.

The tower was built at the height of postmodernism in architecture during what has been called an age of irony. Yet it is hard to imagine that the historicism of Trump Tower was ever meant in an ironic way or that the references to grandeur and gigantism should be taken at anything but face value. Back

towertotower From the atrium in the Trump Tower #pinkmarble #comfortfood #publicspace #AdaHuxtable

in the 1980s, Ada Huxtable was furious that a quotation in which she spoke favorably of Trump Tower's architecture—and that she felt had been taken out of context—was put on display in the atrium. I look for the quotation but can't locate it anywhere. "I'll make a deal, since he likes them," Huxtable wrote in a statement about the wrongful use of her words, a statement that veered between the prophetic and the trivial: Trump could keep the quotation if he refrained from building on Madison Avenue, she offered. Today, it seems like wishful thinking that Donald J. Trump might have stuck to his previous career as a hard-ass exploitative building mogul, directing his energy toward cluttering even Manhattan's most prominent addresses with womby commercial spaces that offer comfort food to people who would never get the chance to dine in a penthouse apartment atop the dark, impenetrable tower hovering above them.

Addressing the question of what constitutes publicness in light of latent gigantism, in this chapter we look at the role of the material setting of urban space, in cities as well as in digital media, and we discuss what defines its constituents in a situation where the usual dichotomous relations between public and private need to be reevaluated.

—HS

The Fifth Leak

On April 14, 2017, the hacking tool EternalBlue was leaked to the public by a group known as the Shadow Brokers. It was the group's fifth leak, and it called this leak Lost in Translation. A so-called exploit used to weaponize a software vulnerability in Microsoft's Windows operating system, EternalBlue had allegedly been developed by the National Security Agency (NSA). The tool takes advantage of a vulnerability in a transport protocol that allows Windows machines to communicate with each other and other devices. Once attackers have established themselves on one device, they can fan out across a network. The impact of the exploit was global and excessive to a degree that made it gigantic, with over 200,000 machines infected over the first two weeks.[1]

The best-known use of this tool is the worldwide ransomware attack WannaCry, which was tracked back to North Korean government hackers.[2] But EternalBlue has also been used to install cryptocurrency miners on target devices and is reportedly used by hackers across the globe, from the

Iran-based cyberespionage group Chafer, which targets personal information from telecommunication companies and the travel industry, to the Russian group Fancy Bear, which has been linked to interference in the 2016 US presidential election.[3]

Yet the spread of the use of EternalBlue is characterized by latency, and latency is key to how such attacks and leaks materialize. Several weeks before the leak, Windows had already released so-called patches to counter the vulnerability. But whereas home office machines often automatically install such updates in the background, corporate and institutional networks often install them after a delay because they first need to test how the updates will run together with the institution's internal digital systems and networks. These temporal irregularities are what made it possible for EternalBlue to enable numerous attacks long after the leak had been patched.[4] Moreover, the name itself, *EternalBlue*, with its emphasis on eternity, creates a conflation of past, present, and future into the color blue, like The Blue Marble: an abstraction, or flattening. However, in being leaked, the exploit made apparent the material reality of institutions that use legacy software and hardware (meaning a system that is out of date but still in use), including healthcare, education, and government institutions, which showed themselves particularly vulnerable to these types of attack, emphasizing an intricate relationship between temporal and spatial latencies.

In the previous chapter, we engage with One World Trade Center not only as a gigantic phallic tower but also as a womb that offers many views—into the city, into history, into digital media representations of these views—and we see conflicting forms of gigantism at work. In this chapter, we explore the surrounding topographies—physical, digital, and cultural—in which this tower is embedded in horizontal terms, the wider container qualities of the area at the foot of the gigantic tower, and the ways the calm embedment of technologies here impacts our experience and understanding of those topographies. In this way, we tackle other aspects of latent gigantism as part of civic urban culture—not so much by focusing on the many surveillance technologies that permeate the area (from CCTV cameras to more sophisticated, automated technologies such as facial recognition) but rather by investigating the relationship between the material and the digital aspects of the memorial culture at the site and the way it can be understood as imprinted by a form of latent gigantism that blurs the architectural and the digital, the material and the technological, in the way our discussion of EternalBlue emphasizes.

The more openly accessible and less immediately mediatized atmosphere of the memorial space at the foot of the tower calls on tropes of civic and urban life, emphasizing that what is at work here is a commonality where nothing is neutral. This has bearings on the imaginaries of commonality evoked at the site and on the ways in which this commonality is spun in a digital web. To access this discussion, we return to the question of how to understand the relationships between media, materiality, and cultural understandings. This question in turn taps into the discussion of relationality broached in chapter 2 in relation to Donna Haraway and Michel de Certeau. Here, we turn to Bruno Latour, one of today's most influential philosophers of relational thinking, whom we briefly encountered in the previous chapter.

We focus on Latour's comments on the six terror attacks in Paris that took place on November 13, 2015. This bridges chapter 1's discussion of the social media uses of the Eiffel Tower as a symbol of solidarity with this chapter's continued exploration of latent gigantism on the Ground Zero memorial site in relation to commonality and public space. Latour's 2015 comments politicized the effects of entanglements between the human and nonhuman in a way that invites us to revisit notions of the civic and urban culture. This is relevant not only for understanding the implications of the fractured topography of terror on that night in November 2015, or for debating about urban space and terror in Paris, but also in relation to the Ground Zero site as it has materialized as a popular destination in the city today.

This discussion offers an inroad to exploring the difficult and loaded question of what common ground entails in the early twenty-first century in light of our discussion of latent gigantism. What is common when deep-seated dependencies between the architectural context and digital media help to form the spatial and temporal structures in which we move and operate? What happens to public space? To commonality? Such questions, we argue, can be illuminated by discussing the impact of the form of gigantism whose seemingly latent presence is marked by spatial distribution and temporal overlay at the intersection between built environment and digital infrastructures—focusing here not on the transmitting tower rising above but on leaks on the ground.

Never Ever Modern

On November 22, 2015, nine days after the Paris attacks, Latour published an opinion piece titled "The Other State of Emergency." It compares the crisis brought about by the terrorist attacks to the climate change crisis discussed at COP21, the 2015 United Nations Climate Change Conference that was to commence in Paris later that month. It is relevant to include here because it links our discussion of relational utopianism in cultural theory with questions of the possibility for commonality in cities in light of terrorist attacks. In this text, Latour considers the attacks to be a foreshadowing of the inevitable failure of the conference and initiatives like it. He has no hope that the peoples of the world will come together to save each other from the climate changes they have themselves coproduced and whose effects they are now experiencing (for example, in the sharp rise in extreme climatic events, including high temperatures, increased precipitation, and storms), and he regards that failure as intrinsic to the political system and the culture it fosters. He even goes so far as to compare the terrorists—to whom he ascribes a blind nihilism—to the politicians who will soon convene in Paris and who he thinks fail to address the issue of global climate change appropriately. He writes:

> Global warming threatens all states in every way: from industrial production, business and housing to culture and the arts. It threatens our values at the deepest level. Here is where states are actually at war with each other, battling for market share and economic development, not to mention the soft power of culture. And each of us feels divided against ourselves. If indeed there exists a "clash of civilisations," then this is it, and it concerns each and every one of us. Yet, we know that national governments are just as lost and helpless here as they are when facing the terrorist threat. The Police aren't enough. Rather civil society as a whole has to take its fate into its own hands and compel political institutions to find answers.[5]

Shortly after the attacks, then French president François Hollande declared a state of emergency that remained in effect for more than six months, the longest state of emergency in France in over fifty years. Latour sharply criticizes Hollande's decision in this text, and we may compare his use of current political events in a larger theoretical argument with that of Jean Baudrillard's post-9/11 comments. Baudrillard noted that there was symbolic power invested in the attacks, to the point where they became an inverted marker of the failure of Western culture, an interior destructive force in which he did not seem to implicate himself. But Latour does implicate himself and all of humanity in the schism that the climate crisis evokes: he talks about a fundamental divide *"within ourselves,"*[6] echoing his own view of the cultural and the material as fundamentally codependent. This codependence does not concern a deep psychological resemblance or parallel but invokes a gigantic expanded field between the human and the nonhuman, an ever-expanding field of relations. While Baudrillard ascribes agency to the material context by anthropomorphizing the Twin Towers, Latour sees agency in the material world more widely: humans and the material world are relationally produced on a scale of gigantic proportions.

As Latour notes, the state of emergency as a political reality in France after the November 2015 attacks puts civil society out of the picture. Latour's calling on concepts of civil resistance hinges on the tradition of political uprisings, even revolutions, in France. It chimes well with activist groups in Paris at the time that attempted to build inflatable cobblestones in their quest for ways of demonstrating that would circumvent the restrictions of the state of emergency, playing on the situationist idea that became a slogan of the May 1968 protests in Paris: "Beneath the cobblestones, the beach."[7] It is a way of calling on a view of public space as a place where resistance can form and make a difference in relation to the more severe crisis that Latour diagnoses—namely, climate change. At Ground Zero, the Occupy Wall Street movement, which had its base on a small strip of public space close by, indicated a similar belief in the possibility of eruptive change. Latour calls for an*other* state of emergency in response to the political one, and he sees it as an occasion to invent demonstrations more innovative than yet "another march from Place de la République to Place de la Nation," as he writes. As we argue, to apprehend, like Latour, civicism through the tropes of revolution and activism means to draw on concepts that pertain to a modern conception of time as moving forward and a dialectical notion of the progression of the world. If the forms of civicism that Latour evokes hinge on such utopian notions, however, we also can argue that they hinge on a conception of public space as a neutral playing field, an understanding that has long been heavily contested in particular by Marxist, feminist, queer, and decolonial theorists but that is also exemplified continuously by online monitoring and datafication of our everyday actions for commercial as much as intelligence purposes.

By comparing the disasters brought on by particular and situated terror attacks to the more elongated, larger-scale ecological processes implied in climate change, moreover, Latour arguably obliterates the ontological differences between them. In Latour's argument, the greater crisis of climate change potentially endangers all of humanity together with other life forms, in contrast to the specific and targeted destruction of human culture and politics inflicted, for example, by terrorists. He refers to this greater crisis as a collective form of human suicide and in this way is curiously close to the Baudrillardian argument about the Twin Towers' so-called suicidal self-destruction. For Latour, industrial culture's role in producing climate change presents a form of collective suicide that implicates all life forms. Latour's version entails in this way a radical relationality between human and other, nature and culture, in relation to which humankind bears moral responsibility, an Anthropocene guilt complex that he relates to all of humanity, despite the gigantic flattening at work in ascribing guilt to "humankind" as all equally responsible.[8]

Moving on from this Latourian position, we therefore aim to ask a different set of questions: What happens if instead of hybrids, entanglements, or assemblages, we look at leaks, stickiness, traps, and effects of parallelism in the relationship between modern material culture and reflexive thinking as such? Archeologist Ian Hodder offers an inroad for problematizing this situation, pointing toward a more cautious path that recognizes the historical conditioning of such theoretical positionings:

> There are problems with the idea of a total mixing of humans and things in networks or meshes. At certain historical moments and in certain contexts, humans appear dominant over things, but at other places and times things seem to have the dominant hand (for example, during global warming at the end of the Pleistocene and perhaps during our own current experience of global warming). In ANT [actor-network theory] everything is relational and this insight is important. But it is also the case that materials and objects have affordances that are continuous from context to context. These material possibilities (whether instantiated or not) create potentials and constraints. So rather than talk of things and humans in meshworks or networks of interconnections, it seems more accurate to talk of a dialectical tension of dependence and dependency that is historically contingent. We seem caught; humans and things are stuck to each other. Rather than focusing on the web as a network, we can see it as a sticky entrapment.[9]

The idea of sticky entrapment, which Hodder offers, coupled with an attention to the ethical implications of this stickiness as introduced by Ahmed, provides us with a structure of difference that takes into account the affordances of the material context and thereby also its differences and dependencies. Indeed, movements such as #metoo and #BlackLivesMatter as well as contemporary feminist, decolonial, and crip theory have made apparent the degree to which technological and architectural affordances are gendered, racialized, and dependent on bodily and mental abilities, as well as the necessity of addressing them.[10] Although what is latent cannot be laid out or expressed, it is nonetheless *felt* by the people involved, and we can discern its effects, for example, in entrapments or leaks that reflect affective responses that are difficult to categorize or even uncomfortable because of the way they stick to us. Equally, the homogenization that it was once hoped global networked communication would bring about here emerges as a form of gigantism that runs the risk of glossing over dependencies and relations between things that are hierarchical and that can lead us to ignore pressing unsustainabilities and inequalities and slide into utopianisms.

With a critical appreciation of Latour's call for civicism and for an*other* political culture that responds to the lack of binary division between the material and the cultural, we can thus begin to tackle the way notions of public space or even the civic take form at the ground beneath One World Trade Center and the different formations of space and time that are found there. What would civicism mean in "another state of emergency" if we extrapolate the latter more widely as a cultural situation where understandings of public and private, materiality and meaning, are being calibrated anew? And how is this understood as a latent form of gigantism? For us, the question of how architectures and digital cultures are held in common (with all the ethical and political issues this entails) as structures of the gigantic and are thereby implicated in forms of civic life is critical. To enter this discussion, we now turn again to Ground Zero at the foot of One World Trade Center.

Calculated Publics at the Edge of the Memorial Pools

One World Trade Center stands on the northwestern corner of the Ground Zero site, which has been turned into a memorial park. The site is laid out as a granite field punctuated by 250 oak trees, with dramatic water-filled pools at the sites where the Twin Towers once stood. The plaza is an open space that can be visited free of charge by anyone in Manhattan. As an open

World Trade Center Memorial and Museum, August 18, 2016. © Paul Sableman.

space in the city, the area is loaded with affective meaning in a way that appears surprisingly unsecuritized, although it is in fact filled with security and surveillance measures visible in the heavy policing of the site, in the large number of private security personnel working here, and also in subtly visible surveillance technology (including facial recognition systems) that reportedly uses AI cameras that can detect unusual movements.[11]

It is also devoid of the media interventions witnessed in the interior spaces of One World Trade Center (discussed in the previous chapter). The decision to build the memorial emerged from the ethical imperative to keep the site sacred in honor of the many people who literally vanished there. Unrecoverable from the debris, their bodies were reduced to the same status as the building rubble. If the site at first glance looks like an artificial forest or grove, we ought to remind ourselves that it is placed on a thin crust: the ground beneath it partly shields the hollow space that holds the memorial museum as an intrinsic part of the site, and it is a largely urban and very dense site that is heavily securitized. Surprisingly, perhaps, despite its banal but also friendly and folksy appearance in drawings visualizing what kind of horizon for gathering people the green setting

National September 11 Memorial. © PWP Landscape Architecture.

would build, the public space in front of One World Trade Center is highly metaphorical and tries hard to call on the highest common denominator of meaning in the commemoration of the approximately three thousand people who died there.

The Ground Zero Site 155

National September 11 Memorial. © PWP Landscape Architecture.

The garden of trees was created by landscape architect Peter Walker. The 250 white oaks planted here seem almost identical, distributed on the site according to a precisely measured grid. Planted to mark the tenth anniversary of 9/11 in 2011, the trees have an architectural quality. Once they

Facade detail of the World Trade Center Building, New York City, in July 1974. © CeeGee, 2015. Wikimedia Commons.

have grown to full size, the trees will form a canopy of leaves that will change color and structure throughout the seasons of the year. The shape and formation of these oak trees will become an upward movement of leaves, branches, and trunks, not unlike the neo-Gothic ethos of the original Twin Towers design. Indeed, the landscape architect himself makes this connection visible and calls the forest a sacred zone. As a uniform entity, the trees resemble an organic forest from some angles, while other angles reveal precisely aligned corridors between them. This is a simulated natural environment where the trees connote the trope of difference in sameness, forming a literal passageway through which one must pass when visiting the memorial proper. Building on a duality between city and nature, the group of trees can thus be seen to form a ritualistic *other* that bears connotations of nature's capacity to transform or even cleanse.

This reading is supported by the way the organically growing oaks embody the fecundity of nature and can be seen as an affirmation of life. As

The Ground Zero Site

The forest. © PWP Landscape Architecture.

the landscape architect emphasizes, the "broad scope of the trauma of 9/11 requires that the memorial use a symbolic language understood by a diverse audience."[12] In the more general narrative of Ground Zero as a memorial space, this is played out against the story of one tree that is important in the branding of the site as an attraction for visitors: the survivor tree, which dates back to before 2001 and which literally endured the attack as well as a later episode when it was removed from the Ground Zero site during the reconstruction phase but was then struck by lightning. This endows the planting of trees on this site, which used to be a much more open paved plaza, with a sense of continuity.

Traditional tropes of gardening mean that deciduous trees' cyclical rhythms point to a sense of eternal renewal, and this comes to comprise an important part of the metaphorical vocabulary used to give voice to the singular event the site marks. Peter Walker's description of the trees' design emphasizes the double function of the site—a place for public commemoration of one particular event in the past and for the more fleeting habitation of the space in the present—by differentiating between September

11 and "typical days": "A grassy clearing within the grove is a quiet space away from the bustle of the plaza. Designed to accommodate ceremonies—specifically, the reading of victims' names annually on September 11—the space also provides soft green park space on typical days."[13] The grove of uniform trees in this way evokes images of community, difference, new life, and a life beyond human finitude through the superhuman forces of nature. These well-known tropes are strongly present at the site but are complemented by other regimes of signification in the symbolic vocabulary of the pools themselves, as we shall see.

The central feature of the design scheme is the memorial pools, which mark the footprint of the Twin Towers. Water flowing across the black granite signifies the way the site is marked by eternity and eruption at the same time—by time as flow and time as event. The centerpiece of architect Michael Arad's design for Ground Zero, titled *Reflective Absence*, is the pair of huge fountains. Each 192-by-192-foot pool is sunk thirty feet into the ground, and their reflective properties are seen in a play with the large flat bodies of the buildings in the background in Maria Finn's drawing that opens this chapter. The stark metaphorical play of the architecture puts on display forces of nature that can be heard, felt, and understood most strongly in the gravitational spectacle of falling water, directly referencing the gravitational pull of the destructive power that led to the deaths of most of the victims and that was illustrated by the pictures of people jumping from the towers.[14] The ten-meter fall of water is a direct and acutely available reference to that free fall.

As an insertion into the site, however, the monument is first and foremost a barrier to people's entry onto the footprint of the towers. Any attempt to climb over the sides of the pool is blocked by a low wall, and climbing over it would be dangerous. This evokes an opposition between nature and culture: what gets bracketed as the unnatural forces of the attacks here succumbs to the realm of the natural, while the monument offers solace in the taming of those forces. There is a reflection of absence, in the literal sense that the properties of water and granite stone (very different than the aluminum and glass surface of the original Twin Towers) in the shape of the towers' footprint remind us of the towers' discontinued existence as well as of the absence of the people who died at the scene. As with the symbolism of the Eiffel Tower when it opened as a testament to the prowess of industrial building techniques at the World's Fair in 1889, the symbolism of these pools as markers of the absent towers is fairly simple and straightforward. But whereas in the case of the Eiffel Tower the motivation of the design by historical events afforded a significatory openness, the anchoring of this monument in the particular event of September 11 gives less symbolic flexibility. It is difficult to imagine this site becoming a lightning rod for meaning, to borrow Barthes's description of the Eiffel Tower, although it is already a tourist magnet to the same degree as the Eiffel Tower and as such is public property. It is as if the singularity of the event the pools commemorate bars them from significatory free play.

The most immediate manifestation of the anchoring of shared memory and individual loss in relation to the terrorist attack is the engraving of victims' names onto the bronze panels that line the shiny granite surface around the pools. The memorial pools and the area at the foot of One World Trade Center on the Ground Zero site are thus simultaneously overlaid with heavy symbolic meaning, a commemorative function of individual and collective grief, and a poignant aesthetic form. The 2,983 names on the bronze panels are not organized alphabetically or in chronological order but according to an algorithm that was fed with the complex interpersonal relationships of the victims. The algorithm designated these relationships as meaningful adjacencies, mapping coworkers, friends, family members, and other reported social linkages. The algorithm was purpose-built to handle the layout: first clustering the names, then layering them into the overall design. The order is not discernible but marks an abstraction of social relations that speaks to the flattening of the horizon discussed in chapter 1 and at the end of chapter 3, pointing to a publicness that is not only networked but algorithmically calculated.[15]

The notion of *networked publics* has become seminal terminology in media theory to describe the collective that results from what media theorist danah boyd calls "the intersection of people, technology, and practice."[16] In this line of thinking, publics are made up of users who assemble through social media, but the concept also entails an understanding of publics as conditioned by the affordances that technology offers—that is, the technological conditions provided for those users. By addressing technological affordances, we are not presupposing a predigital, unmediated mode of interaction but are stressing the kinship between technological affordances and the architectural affordances given by the built environment. The towers we deal with in this book are both media and materiality. As we have emphasized throughout the book, to interact face to face is no less culturally inflected than to interact digitally. Communication is always mediated in complex ways, which is why it is important to consider more precisely how it is embedded in concrete material and situated environments such as the memorial pools. The fact that an algorithm was employed to lay out the grid of names on the sides of the memorial does not make it any more or less a place for the emotions that the space is meant to host. However, we need to consider what affordances this spatial outline of the names entails for the communities in which they are ingrained: that is, how those emotions are facilitated by the material conditions being offered.

On Maya Lin's famous 1982 Vietnam Veterans Memorial in Washington, DC, the names of soldiers who died in the Vietnam War are organized in chronological order of their loss and chiseled into a black stone wall. The written words are an individual bestowal of meaning in relation to the arguably more immediately recognizable organizational principle of the chronological ordering. At the Ground Zero memorial, the logic of the order of the names is created algorithmically and is not discernible to the human eye. Whereas Maya Lin's monument enables spatial and semantic openness in the way it gently cuts into the field, Michael Arad's monument delegates meaning-making to the calm workings of the digital realm. This enables a sense of latency that is put into effect by the overlaying of different temporal registers, ingrained in spatial categories that are simultaneously

World Trade Center site. © Studio Daniel Libeskind.

and indistinguishably material, mediated, and calm, configured digitally as much as architecturally. We see this in the way the algorithm has been programmed to take account of who worked in the same offices and who interacted with whom.

The algorithmic calculation of these meaningful adjacencies invisibly engraves tales of heroism and compassion in the form of small spatial stories that took place inside the now-collapsed towers. An example is the story of two office workers who met on the staircase while trying to escape and died in the same place. They had apparently never met before that day, but their names now sit next to each other on the memorial. Such stories of social and physical adjacency across a range of different timespans—from brothers who grew up together to people who got to know each other only while trying to escape—remain latently present in the memorial as an abstract articulation that conflates temporality with the *now* of the aftermath and the commemoration to which the memorial gives shape. Whereas algorithms that calculate publics are often employed to predict and preempt consumer patterns or security risks, the algorithmically calculated public formed by the victims' names at the Ground Zero memorial gives shape to a temporal conflation and a form of meaning-making that is both gigantic and latent at the same time. By mapping relations between victims but leaving the nature of those adjacencies open to speculation, the memorial neither freezes the past nor projects a fixed narrative into the future. Its gigantism remains latent, hinging on the interplay between the calculable and the incalculable.

A consideration of the algorithmic embedment of the built environment and of the spatial and temporal properties this embedment affords points to a latent gigantism that comes from this flattening of meaningful adjacencies into smooth bronze plaques. The burning question is: what makes something meaningful in this context? With the bronze plaques on the memorial pools, the algorithm that generated the order of the names is invisibly present in the display. Yet the plaques do not show other latently present stowaways, such as the undocumented victims. Their names are not visible in the commemorative space because officially they were not there, meaning that they cannot be calculated as part of the public the algorithm configures. This raises questions about who is included, who is excluded, and how technologies constitute and codify the very publicness they identify—what media theorist Tarleton Gillespie calls *calculated publics*.[17] This term points to the fact that such publics are often constructed in opaque ways—for instance, when Amazon or Netflix recommends material based on the previous actions of users like you.[18] These companies generate communities whose constituents are not transparent (because the algorithms are trade secrets) but that are potentially gigantic, insofar as algorithmically calculated publics are part of the conception of the collectives in which we all participate. As Wendy Hui Kyong Chun argues: "As characters, we are never singular, but singular-plural; I am YOU."[19] It is this plurality that identifies probable actions and collective habits. The 2016 US presidential election highlighted the risk of such calculated publics and of the so-called filter bubbles and echo chambers that may accompany them.[20]

The discussions surrounding the idea of calculated publics allow us to understand more fully the way in which a sense of publicness is articulated at the foot of One World Trade Center, where the commemorative space takes on material form in the names inscribed around the pools. These discussions also ultimately allow us to understand how the gigantic is latently embedded here. Although the algorithmic calculation of the included names' spatial outline remains out of sight, it is part of the physical experience of the site. The site calls attention to its own calmly present organizational principle precisely through the way this order is *not* immediately identifiable to the human eye, as, for instance, an alphabetical ordering would have been. In this way, the monument gestures toward the organizational techniques that structure so much of our public engagement today, and with which the advent of scandals and leaks reveal the gigantism of their range and influence.

The notion of *calm technology*—a term suggested by computer scientist Mark Weiser (1952–1999) of the Palo Alto–based research and development laboratory Xerox PARC—epitomizes this gigantism. As one of the founding researchers into ubiquitous computing in the early 1990s, he used the word *calm* to describe technologies that integrate seamlessly into our everyday lives without calling attention to themselves: "The most profound technologies are those that disappear. They weave themselves into the fabric of everyday life until they are indistinguishable from it," he wrote in 1991 in a short text titled "The Computer for the 21st Century."[21] At that time, Weiser and his colleague John Seely Brown saw the emergence of the internet as establishing a paradigm of distributed computing on a large scale. They envisioned that this would increasingly decouple computing tasks, as well as interactions with the computer as object, from the precisely defined machine on which an individual depends. This would mark a shift to a networked technology that would be distributed across many platforms and large distances yet also be accessible from particular points in space. Weiser predicted that distributed computing would eventually be followed by what he and Brown termed *ubiquitous computing*. Ubiquitous computing, in their words, is "characterized by the connection of things in the world with computation," which "will take place at many scales, including the microscopic."[22] Regardless of the scale at which they functioned, these connecting technologies would operate in a way that was largely invisible to the user because they would be either hidden from view, very small, or spread over large distances. Through their ubiquitous, integrated, and semiautomatic functioning, such technologies would recede into the background of everyday life—hence their characterization as calm.

Weiser and Brown's understandings of the formation and effects of ubiquitous computing were predictions rather than observations, but their writings come surprisingly close to describing the design impetus of many of the so-called smart technologies and the emergence of the internet of things.[23] Such technologies may be found in the myriads of information-collecting chips and processors that are integrated into the physical stuff with which we are surrounded on an everyday basis—chips that are often in some level of intercommunication with one another, not just in our smartphones or computers but increasingly also in furniture (the fridges of "smart" home technologies that signal when the family is running out of milk), fittings (systems that shut off the house's main water pipe if there is a leak), and architecture (embedded sensors that register comings and goings in the office). Calmness, with its strong connotations of neutrality, remains a design ambition in much of the research that goes into the development of artificial intelligence and machine learning, and it is related to the rhetoric that big data is data that speaks for itself.[24] Yet there is something more at play than the metaphor indicates. As critical data studies have shown, biases and inequalities persist, and this requires us to pay attention to the gendered, racialized, and ableist presumptions that are carried over and even allowed to multiply: we must address the latent gigantism at stake in taking the global, homogenizing vocabulary at face value.[25] This requires an attention to materiality and to dependencies rather than flattenings between nature and culture, human and machine.

Although information and communication technologies are invisibly present in some places, they emerge in more tangible form in other places where digital labor practices, energy politics, server farms, and data centers materialize and take physical form. They emerge in areas that provide particularly beneficial circumstances for those functions—whether in the availability of cheap labor or a favorable climate for keeping servers cool. Geographical, social, and political infrastructures have not vanished with the emergence of the digital realm but have been rearranged in a seemingly smooth global geography with local impacts. The interrelatedness between the materiality of technology and its cultural meaningfulness becomes most apparent when a disruption occurs behind sleek interfaces that are pleasant to touch—the tangled wires, the looping malfunctions, the leaks. Even if location and geography are massively stretched out to span the entire globe, they may not necessarily be "global" in the sense that they can form a uniform or even stable horizon for meaning.

From Containment to Leaks

In his investigation of the grid as a cultural technique, German media theorist Bernhard Siegert relates: "For Mies van der Rohe only a skyscraper under construction was a real skyscraper, for only so long as its sides had not been closed and covered was the steel skeleton able to make the constructive idea transparent. Glass facades were therefore for Mies no more than a compromise with the inevitable."[26] The Eiffel Tower is in this sense a real skyscraper since it makes the constructive idea transparent. As we have seen, the Eiffel Tower is a completely transparent structure without the reflective qualities of glass: it lays bare its own construction and has no walls, except in the small laboratory at its top. As a vantage point, the tower provides an overview of the city of Paris; however, as we have shown, the trip to the top breaks up this view into a montage of pointed angles, making it fluid and adaptable. We have also seen the Twin Towers as a container of people, information, and wealth—all things that characterize a transition from industrial to information society—and boasting an antenna at the top that took the TV signals and cell phone communication lines of thousands of New Yorkers with it when it was destroyed. The facades of the Twin Towers, with their narrow windows and strong vertical lines, connoted solidity and stability. Transparency here was more a question of looking out than of looking in, giving an impression of light

One World Trade Center and Seven World Trade Center at night, viewed from the 9/11 Memorial's plaza. © Antony-22. Wikimedia Commons.

that leaked horizontally through the bars, disrupting the vertical lines and revealing the semblance of an interior world.

Finally, we have seen the One World Trade Center tower as a repository that brings together the various ways we can talk about containment today—both as collections of people assembled in a dense vertical container (a womb with a view) and as digitally mediated experiences (such as the elevator's mediation of the history of New York, a conflation of time and space inside a moving container). This is a building that is simultaneously wired through and through and permeated by wireless electromagnetic signals. It displays an ambiguous attitude toward transparency—from the mock performative gesture of transparency on the lower securitized ground levels, to the upper levels that in sunlight appear like a gigantic mirror reflecting the city's surface, to the full-on mediated performance of transparency on the observation deck. This transparency is a question not only of glass but also of screens, and it reveals itself as an ideological and spectral category that is linked to modernist ideologies and takes curious forms in the twenty-first century. The building presents itself as securitizing itself against leaks as well as insertion, yet the enormous antenna reminds us of the ways the building is both wired and wirelessly connected in networks that leak. This speaks to a broadening of public space from the physical cityscape to the potentially gigantic dimensions of possible digital encounters facilitated by the wires and wireless signals that envelop us in an infrastructural grid, of which the antenna atop One World Trade Center is also part. In recent years, the most marked aspect of this grid has become social media as leaking containers of data ripe for analysis by marketers and politicians alike.

If we descend into the ground beneath Ground Zero, we find a very concrete and material example of containment and leakiness and their implications. In the 9/11 Memorial Museum, part of the slurry wall around the World Trade Center foundation that holds back the pressure from the muddy ground next to the Hudson River is exposed. This ninety-one-centimeter-thick concrete structure prevented the Hudson from flooding parts of the subway system when the towers fell, and it has subsequently become a symbol of resilience in the extensive mythology of 9/11.[27] In a *New York Times* article from 1966, when the foundation was laid, the slurry wall was explained as "a giant empty bathtub."[28] If the large pit necessary for the foundation and underground areas of the towers had been dug without being secured in this way, it would quickly have filled with water: the bedrock is sixty feet below ground, while the water table is only five feet below ground.

The slurry technique, which was employed for the first time in the United States by architect Minoru Yamasaki's team when it built the World Trade

The Ground Zero Site

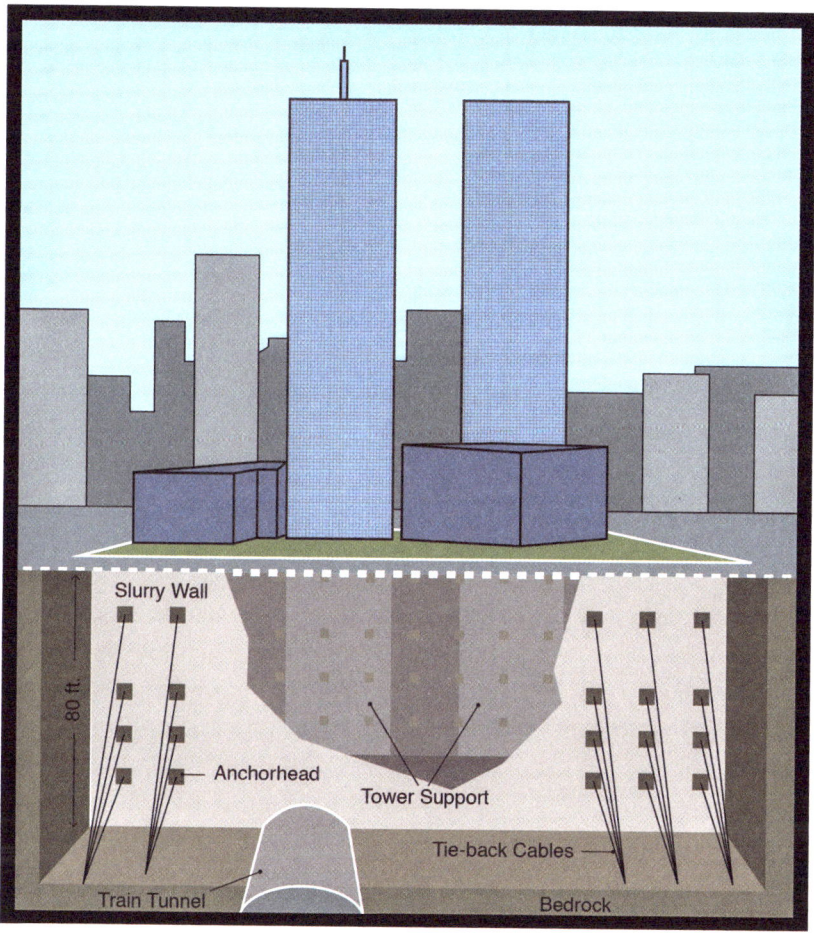

Slurry wall explained. © Kristen Van Haeren.

Center, consists in digging a deep, narrow trench down to the bedrock. The trench is then filled with slurry, a special powdered clay containing the mineral bentonite. Bentonite takes on a thick, gooey consistency when saturated with water. It coats the sides of the trench and has the ability to plug a leak if one should occur. Then concrete is poured in from the bottom, pushing the slurry upward, displacing it and forming a waterproof wall, which is subsequently anchored in the bedrock. As we discuss in chapter 3, the large amount of material that was cleared before making this slurry wall was used to create the artificial land that is now Battery Park. This essential part of the construction of the towers was never meant to be seen but was to remain situated in the ground. Only in the massive disruptive event of the collapse of the towers was it laid bare.

In Daniel Libeskind's original design for the Ground Zero site, part of the slurry wall was retained as a significant design element and a testament to what did not crumble in the attacks. This feature has been kept in the Memorial Museum, where part of the slurry wall is exposed. However, when the museum was still under construction in 2012, New York was hit by Hurricane Sandy, which caused the Hudson to rise and spill over into the pit, flooding the museum space and also the slurry wall from the inside. The disruptive event of the terror attacks had not made the slurry wall leak when the towers collapsed, but the hurricane rendered the construction vulnerable in a new way. In the slurry wall, we see how the notion of leakiness embodies the properties of the spatiotemporal overlays we are dealing with here. The slurry is a composite fluid material that seeps in and performs an important function in securitizing against collapse but it is eventually displaced by more muted material that renders its essential contribution invisible. The wall also forms a curious dialogue with the memorial pools' concave structures. The slurry wall was meant to keep out the water table, which threatened to fill up the construction hole with water from the Hudson. In the memorial pools, water flows in a controlled and continuous manner, keeping a human audience away from the sacred space of the footprints where the towers used to stand.

Beneath the slurry wall bathtub, we find hundred-year-old train tunnels, signifying the temporal linearity for which the railway system would be a fitting allegory. In the restructuring of the Ground Zero site, the adjacent train station has also been redesigned. It is modeled to look like the inside of what looks most of all like a gigantic whale skeleton, framing the technology of the train (emblematic of industrial progress and the American frontier spirit) as a relic of former glory, or even as reminiscent of the filigree ironwork of structures like the Eiffel Tower. In this way, it engages in a curious dialogue with the museum commemorating the collapsed Twin Towers. Framing the site as an archeological excavation of the immediate past, the structure plays with notions of temporality that are not dissimilar to those of the train station, uncovering a future experience of the excavation of our present. This condenses the twentieth century into a field of simultaneity located in a future when the gigantic bodies of whales and towers might

A portion of the "bathtub" slurry wall from the World Trade Center as seen from above in the National September 11 Museum at 180 Greenwich Street in the Financial District of Manhattan, New York City, 2016. © Beyond My Ken, 2016. Wikimedia Commons.

have succumbed to the passing of time and been made extinct so that they only remain as monuments to themselves—on the one hand in the form of memorials, and on the other as almost mock versions of their own late nineteenth-century ambitions—even while the waters are rising.

If President Donald J. Trump, whom we encounter in the prologue, is an example of a nostalgic longing for the greatest or tallest, this longing concerns particular forms of gigantism that have to do with ideas of progression and production. Almost paradoxically, however, his power arguably resides in a latent gigantism that is crudely visible in his constant Twitter comments (which can infuriate, entice, or create a sense of blasé overload in the reader) *and* at the same time is latently powerful (as in the ability to solicit the leaking data available online to target supporters or mute opponents on social media or in speeches).²⁹ Commonality and calculated publics, which have been at the heart of this chapter, are thus operationalized here in a way that reveals their ambiguous status in the early twenty-first century. When invisible workings partly orient the spatial and temporal topographies in which we move and from which we get to know the world, it is possible to approach the question of what is common through particular situated readings of digital and architectural spaces that are all part of longer historical trajectories—trajectories in which the human (or in the first instance, perhaps, the reflexive human subject of modern Western culture) may be only an interlude.

If transparency, as a way of breaking open the container, is central to the dominant paradigm of skyscraper aesthetics across the twentieth century, as Siegert argues, then leaks may be seen as a condition of the contemporary spatiotemporal configurations we are looking at here. In the same year that the Twin Towers were attacked, surveillance studies scholar David Lyon described how data moves freely between different sectors of society, resulting in information from discrete realms spilling over into other contexts. He described private life, work life, and shopping as "leaky containers,"³⁰ thereby anticipating more recent arguments in media studies that see new media as leaky by nature. Leaking (as we use it in this book) challenges the promise of neutrality of large infrastructure in a way that both breaches and sustains the boundaries between private and public. Leakiness materializes in digital information transfer and is vulnerable to exploitation, as we saw with the EternalBlue hacking tool.

Moreover, exploit tools designed for specific, targeted insertions can themselves be leaked to a wider public, propelling their reach and impact onto a global and indeed gigantic scale. If we look at security companies that live off securing institutions and companies against such attacks, we see a constellation of fear of material damage, fear of invisible attackers, and uncertainty about when the attack happened as opposed to when it will show itself. Still feeding on the fear of EternalBlue, the small Nigerian consulting company Gigasec reported in July 2017 on the successor to WannaCry, dubbed EternalRocks: "So EternalRocks has the potential to spread faster and infect more systems. EternalRocks is currently dormant and isn't doing anything nefarious such as encrypting hard drives. But EternalRocks could be easily weaponized in an instant, making the need for preventive action urgent. . . . EternalRocks does not have a kill-switch which helped curtail WannaCry and mitigate the ransomware damage. The clock is ticking with EternalRocks; contact Gigasec Services to stay protected."[31]

The vocabulary here is telling. The latent gigantism of the threat is hypothesized to be "faster" and "bigger," and the need for urgent response is emphasized: "The clock is ticking." Moreover, in light of our reading of the slurry wall beneath the Twin Towers, the name EternalRocks expresses a contraction in a way that is telling of latent gigantism: *eternal* speaks to a temporality of latency and a conflation of past, present, and future, not unlike the invitation to See Forever in the One World Observatory. In combination with *rock* in the name EternalRocks (which describes new and potentially bigger threats, and which draws on EternalBlue) solidity and durability are stressed. In this way, it points to the themes of ontological flattening and the broad present that we have discussed in relation to latent gigantism.

Leaks—whether the overflow of the Hudson during Hurricane Sandy or the Shadow Brokers' leaking of NSA spying tools—and the varying ways they can be coread provide us with an inroad to address this flattening. As an articulation of the fluid, the leak (like slurry) points to the ontological slippage that is characteristic of latent gigantism, and paying attention to it may provide an inroad for critically engaging with the effects of that form of gigantism.

Maria Finn, *Unfinished #22*, pencil on paper, 29 × 42 cm, 2018. © Maria Finn.

5   Into the Ground: Cataloging Latent Gigantism

Prologue: The Earth Leaks

Copenhagen, March 2019

Since I moved back to Copenhagen after living for six months in Cambridge, Massachusetts, my phone has continued to notify me of the local weather conditions back in Cambridge. I have tried changing the location settings, but the reminder of the Cambridge weather continues. It is a strange roller coaster of seemingly wild temperature shifts. I never quite got used to measuring temperature in Fahrenheit when I was in the States, and when I look at the weather summary feed from Cambridge on my phone every morning, the Fahrenheit scale adds to the drama of changing temperatures: even in February, you get temperatures in the thirties one day, in the tens another, and so on.

These temperature shifts make me anxious. The rift is not just between two places and two time zones, where the digital feed stands in for a place it will be hard to leave behind completely. It feels like an information leak I did not ask for and I wish I could ignore. In the more stable conditions of the wet, windy, cool Danish climate, which is arguably moderated by the Gulf Stream—not unlike the way human frailty is moderated and homogenized into a pale but picture-perfect mass by the infrastructures of the weather-beaten Danish welfare state—the twists and turns of the New England weather seem a strange warning sign of what might come. If global warming and climate change implore us to consider scale relationships that go beyond local contexts—relationships to do with the gigantic scales of the earth, the evolution of life, or even the solar system—this leak reads more like an allegory than a symbol.

Another mismatch of time and place in which I have found myself caught since I returned to Copenhagen is more analogue in character. When I look at my black plastic Swatch wristwatch, with its fluorescent hands and a small box showing the day and date, it is as if not just my phone but also my watch is stuck in another place on the planet. I bought the watch on Amazon as a birthday present for my son, but he said he did not need to wear a wristwatch, so he gladly gave it to me when my old watch broke. In his school in Cambridge,

towertotower
Cambridge, Massachusetts

tuesday TODAY						32	18
Now 0	0	0	0	0	0	0	
☀	☀	☀	☀	☀	⛅	⛅	☀
19°	23°	25°	27°	28°	30°	30°	30°

wednesday	⛅	27	12
thursday	⛅	28	12
friday	☁	37	25
saturday	☀	37	19
sunday	❄	41	39
monday	⛅	48	34

Liked by **mariafinn_** and **1 other**

towertotower Weather app #forecast #timeoutofjoint #planetaryscale Graphics: @Kristen Van Haeren

there was a clock in the classroom. In Copenhagen, he can see St. Peter's baroque church clock tower from his school. While we were traveling back to Copenhagen, we got caught in a snowstorm and missed our connecting flight in Newark. In the hazy mental zone that appeared as we ran through the airport, I changed the time on my watch, and now the date changes at midday rather than midnight. It has been a source of confusion and irritation for weeks. My watch might as well be twelve time zones away. Kristin makes fun of my analogue conflation of space and time, remarking that it is as if my wristwatch had moved to Honolulu while the rest of me moved back to Copenhagen.

I now realize that all it takes to put the watch right is to move the hands twelve hours backwards to synchronize day and night. Like my weather app problem, the confusion is due to a trivial systemic flaw. But if we regard this flaw as an information leak, it comes with a type of gigantism that relates to the scale of the earth. It is a reminder of our earthbound existence rather than a quantitative measurement of longitudinal degrees, feet, Fahrenheit, or the hours it takes to traverse the globe. The situated readings centering on the material context that we have offered in this book unfold the imprint of gigantism today and trace its recent cultural history. They offer a way of grappling with the conditions of earthly habitation that arise at a time when particular interrelations and dependencies among humans, architecture, and technologies are in focus. In this concluding chapter, we continue these situated readings, but we zoom out from our previous sites to consider the wider applicability and implications of latent gigantism.

—HS

Tower to Tower and Beyond

If latent gigantism has a temporality, it cannot be conceived of as one that stretches toward an unknown future with the aim of overtaking what went before. Rather, it can be conceived as a broad moment that has chewed the past into bits that now linger in more or less recognizable forms. Any hope invested in these bits relies on the future *not* becoming what global climate models and simulations predict. Indeed, it relies on the actions we take changing this gloomy course. This temporality makes for an extended present that is immensely large and heavy with responsibility but at the same time endlessly small and futile, depending on whether we are looking at it from the scale of the human or the planet. The climate crisis is a spatiotemporal narrative that is constantly with us, continuously adding new episodes to its horrifying tale. It makes little sense to anticipate an end in the hope that it will ascribe meaning; rather, we may desire endless deferral. Not unlike what happens when we scroll through a newsfeed, the climate crisis narrative operates a temporality of the episodic and speaks to a time of monumental serialization in which each moment spills over into the next.

In chapter 1, media theorist Wendy Hui Kyong Chun reminds us that digital media's logic thrives on crises insofar as they solicit real-time responses, but at the same time it also threatens to exhaust us with repetitive acts of an almost compulsory nature. This means that networked media accustom us to the experience of crisis as a permanent state rather than an exception. We saw this with the symbolic appropriation of the Eiffel Tower and the World Trade Center when terrorist attacks struck. These buildings became symbolic carriers of a sensation of crisis that people can relate to—for instance, by posting images of the Eiffel Tower on Facebook—but that also exhausts itself as disruptive events such as hurricanes, earthquakes, and acts of violence continue to occur across the globe.

As an articulation of crisis, climate change also solicits constant response. As we take upon ourselves the moral authority to act responsibly, those of us who have the resources to act are made complicit in an inverted way, bestowing a sense of individualized responsibility for managing the crisis through concrete actions such as sorting our trash or switching to energy-saving light bulbs. These measures are not unlike the workings of addictive online gameworlds such as FarmVille, which bestow responsibility on their players for keeping the (fictional) world afloat and which chime with the perpetual crisis of networked digital media and the temporality of repetition and succession—from crisis to crisis, from tower to tower.

The climate crisis towers above us as a constant threat to any sense of future, but if we are lucky enough to live in regions not afflicted by life-threatening drought or flooding, it also fades into the background of our everyday lives and can become ingrained for example as mundane chores. We might say that it becomes invisible precisely *because* of the monumentality of its scale. As such, it bears the hallmark of latent gigantism.

One question that remains is how to build for and in light of gigantic scales that remain latently present in our everyday lives. The architecture of gigantism, such as the towers we have discussed in this book, is accustomed to think in temporalities that reach beyond that of the individual architect, and many gigantic or monumental constructions are built in the anticipation that they will see several generations pass through them, sometimes even instilling the cruel optimism of the idea of eternal life on earth. With their transmitting antennas, these buildings remind us of what architectural scholar Mark Wigley calls an antenna ecology, as we quote in the introduction to this book, using this holistic nature metaphor to describe the contours of relationships in which we are embedded and that transgress the scale of the body. The antenna is visible on top of large towers as much as it is invisibly embedded in communication technologies that surround us and that we carry with us everywhere, to a point where they are often seamlessly integrated into our material context and even into our bodies.

As Ara Wilson aptly remarks in her work on the politics and history of infrastructure, the "modern is unthinkable without its infrastructure: without dense housing, transportation arteries, electric power, and now, digital signals. But if it works, what feels new and modern—the technological sublime—eventually becomes part of the background environment—the barely seen telephone pole."[1] Furthermore, the effects of climate change make new demands on constructions to withstand the likely more extreme climatic events and conditions to come, even as they render the futility of building for the future ever more apparent. Like any other human-made structure, these buildings are not perfect or perfectly preserved but are constantly in the process of deterioration, potential destruction, and damage, and in need of repair and maintenance if they are not to end up as ruins. How, then, to understand the paradoxicality of the gigantism of building tall at a juncture when this is by all accounts environmentally unsustainable?[2] How to grapple with a gigantism that is simultaneously more apparent, more embodied, and more latent than ever before?

When we talk about latent gigantism, we are proposing an apparently contradictory conceptual pairing where a form of invisibility (latency) is part and parcel of a form of visibility (gigantism). Cultural theorist Hans Ulrich Gumbrecht has used the term *latency* to argue that the Western twentieth-century conception of time has been transformed: whereas a linear directionality that aimed toward change and progress was once dominating, now a condition of simultaneity that he calls a *broad present*—a present where it is impossible for us to leave the past behind—has taken over. He links this transformation to the post–World War II period, and he analyzes it from the perspective of Germany, where the communist East and capitalist West both tried to leave the past behind and move on, economically as well as morally.[3] Although both states touted narratives of progress toward a brighter future and a break with the past, there was nevertheless something that was impossible to leave behind. As well as situating this sensation through a narrative of his own life, as a representative of the baby-boomer generation, Gumbrecht references Dutch historian Eelco Runia's conception of presence as linked to the stowaway:

> In a situation of latency, when a stowaway is present, we sense that something (or somebody) is there that we cannot grasp or touch—and that this "something" (or somebody) has a material articulation, which means that it (or he, or she) occupies space. We are unable to say where, exactly, our certainty of the presence comes from, nor do we know where, precisely, what is latent is located now. And because we do not know the identity of the latent object or person, we have no guarantee that we would recognise this being if it ever showed itself. Moreover, what is latent may undergo changes while it remains hidden. Stowaways can age, for example. Most importantly: we have no reason to believe—at least no systematic reason—that what has entered a latent state will ever show itself or, conversely, not be forgotten one day.[4]

The stowaway is a figure of containment, and Gumbrecht evokes several containment figures, which he traces in a diverse range of cultural texts from the period after 1945. One notable example—which we mention in our introduction, drawing on cultural theorist Zoë Sofoulis—is philosopher Martin Heidegger's discussion of containers from the postwar period. Gumbrecht calls attention to Heidegger's famous reading of a jug in his 1950 essay "The Thing," where the jug-as-container marks the jug's inner emptiness as a space that is able to gather dimensions and things.[5] Latency in this context establishes that something remains hidden even though it is still arguably on our horizon. But this teetering form of presencing is not necessarily passive or repressed. Sofoulis stresses that Heidegger's analysis of the jug conceives of containment as capable of outpouring and re-sourcing, in the sense that it can carry things over from one source to another and preserve its contents over time.[6] This gives the jug-as-container a temporal dimension that confers leaking properties on it.

As we emphasize in our discussion of latent gigantism, we are not dealing here with the conflation of binary categories such as past and present, visibility and invisibility, containment and leaks, local and global. In chapter 1's discussion of appropriations of the Eiffel Tower on Facebook following the Paris terror attacks in 2015, for example, we emphasize a situation where it was impossible to rely on linear conceptions of space and power. Like the contemporary city itself, the digital sphere of social media constitutes a stage where questions of who can access certain spaces and ascribe meaning to them are negotiated and where contestations and power struggles take on intricate forms of spatial and temporal overlay. This means that despite these infrastructures' allegedly neutral and democratic constitution, they can serve to solidify structures of inequality and preexisting biases. Moreover, as we see in our discussion of the Ground Zero site today, by investigating the effects of latent gigantism on the experience economy of the neoliberal city, we can also call attention to specific opportunities as well as challenges for commonality. Latency therefore is a spatial as well as a temporal category.

Digital technology has enhanced the way the future can be seen as a perpetual part of the present—for example, in the form of a series of risks to be predicted and preempted through data mining and predictive analytics. As we see in our readings of One World Trade Center and its topographical context in chapters 3 and 4, these practices are part and parcel of the design of architecture and infrastructure in twenty-first-century urban culture. Rather than speaking to an ironic significatory play with the past or to dialectical images that force the latent into visibility, the gigantism at work here is simultaneously gigantic, visible, latent, and hidden.

This process of ontological flattening brings with it intricate dependencies between descriptive dichotomies of modern Western culture, such as media-materiality, symbol-function, nature-culture, and humans-technologies. We point to some of these in our discussion of the slurry wall beneath the World Trade Center and in our mediated engagements with the view from the One World Observatory.

From a political perspective, this flattening risks limiting our possibility for knowledge, but it also allows us to bypass dichotomies and understand the intertwinement of apparent oppositions, giving us a way to tap into the embodied dimensions of things. The reading we have offered of some of the formative dichotomies of modern culture resists creating new visible structures, projection planes, or conceptual hybrids and insists on the stickiness of what we are looking at. If latent gigantism means that something remains hidden but is not forgotten or has disappeared, it is urgent to address the dependencies it builds today as ecological and financial crises loom and inequality is on the rise in many parts of the world. We are forced to ask uncomfortable questions about how what remains hidden is not the same for everyone, which in turn undermines the sustainability and legitimacy of the Western reflexive subject's interpretive capabilities as claiming objectivity.

Insofar as this book responds to a series of severe crises at the beginning of the twenty-first century, we have sought to speak about the gigantism we meet in phenomena that cross architecture and digital culture and that play out latently. These phenomena are simultaneously ubiquitously and calmly present and starkly visible in the tall transmitting towers we discuss in this book. We have regarded these towers and their mediated instantiations from multiple perspectives that intertwine and interlink throughout the book: as architectural constructions in their own right; as icons open to cultural projections; as elements in the communication infrastructure of modern societies; and as symptoms of the increasingly precarious and unsustainable foundations of modern industrial production, finance, and political and environmental culture.

They help us to unfold a series of inherent cultural paradoxes concerning why high-rises of gigantic proportions are still being built even though such structures are economically and environmentally unsustainable. The tall buildings we have engaged with in this book can be seen as particularly emblematic examples of this paradox, yet our analysis can be extended to other structures that embody an oscillation between excessive scale and latent invisibility.

As we have traveled from tower to tower, a number of central concepts have emerged as stepping stones for our argument: containment, leaks, flattening, commonality, crisis, and entrapment. These concepts all contribute to an understanding of the dependencies among humans, architecture, and technology—not as assemblages or entanglements that eradicate difference but as spillovers that create flows and dependencies and that require an understanding of their differentiation.

In the remainder of this chapter, we will see these key concepts at work in three short readings of selected sites and phenomena that take us from 9/11 up to today: the China Central Television (CCTV) headquarters building in Beijing, completed in 2012 and designed by Dutch starchitect Rem

Koolhaas; the phenomenon of the tech giants, such as the American company Amazon and the Chinese company Alibaba, and their charismatic CEOs; and large-scale climate management projects in New York, including one by the Danish architectural firm Bjarke Ingels Group (BIG), headed by a former Koolhaas employee, and another to regenerate a former landfill site where large debris from the World Trade Center now rests. These readings challenge the nostalgia attached to the Western ownership of gigantism claimed by the sites we have studied so far, and they point to the wider applicability of what we have identified as latent gigantism.

Kool Chinese Twisted Twin Towers

The year 2001 was not only the year when the World Trade Center was destroyed in a terrorist attack and the situation named the War on Terror by the US administration, legitimizing military aggression, began. It also marked other reconfigurations of the global political scene. It was in 2001, for example, that China formally entered the World Trade Organization, a move that heralded the country's opening up for increasing political, economic, and business cooperation with the West.

This opening was also reflected in the discipline of architecture: there were a number of international design competitions in China involving world-renowned, starchitect-driven projects, such as the 2002 design competition for the CCTV headquarters, known today as the China Media Group (CMG) headquarters, and the 2003 competition for the Beijing Olympic Stadium. For Western architects in the early years of the millennium, however, the idea of building large, prestigious projects in China was still a new domain, riddled with underlying issues about embedded Western values and the massive spending associated with these spectacular buildings, as architectural historian Charlie Q. L. Xue has argued.[7]

As is discussed in chapter 3, 2002 was also the year of the highly publicized and not uncontested competition for the World Trade Center site. The Dutch architectural firm Office for Metropolitan Architecture (OMA), headed by the controversial and controversy-seeking architect Rem Koolhaas, decided to abandon the Ground Zero competition and focus instead on an invitation to enter the CCTV competition in Beijing. Koolhaas had just cemented his fame by winning the 2000 Pritzker Architecture Prize, the highest recognition in the field. His participation in design competitions outside the West might therefore have seemed surprising to some people at the time. In a typical Koolhaasian move, however, he defended his decision with the publication in 2004 of a "Beijing Manifesto" (a counterpoint to his 1978 book *Delirious New York: A Retroactive Manifesto for Manhattan*). In this text, Koolhaas pitches his work in China as an opportunity to bypass the capitalist, consumer-driven context of the "backward-looking US,"[8] where, he argues, there is "no more public." Instead, he is going to "work for the people."[9] He claims that this choice was cemented by the words of a fortune cookie motto he received at a Chinese restaurant during a meal when he was contemplating whether to take on the challenge.[10] As architectural critic Deyan Sudjic suggests, however, it is notable that he made the decision at a time when many of his American projects were facing obstacles and rejection.[11] Nevertheless, the CCTV competition jury included members who must have been sympathetic to Koolhaas's brand, including Japanese architect and fellow Pritzker winner Arata Isozaki and Charles Jencks, the chief American theorist of architectural postmodernism (whom we encounter in chapter 3). They selected OMA's structure—two leaning towers bent 90 degrees at the top and bottom to form a continuous tube, with an uneven, grid-like pattern running across their surface—over, for example, the proposal for an extremely tall skyscraper by Skidmore, Owings, and Merrill, the company that ultimately became the architects behind One World Trade Center.

Koolhaas's twisted building presents a radical reimagination of the twin tower skyscraper idea. The bent, rectilinear form establishes a vacuum in the center that marks out a space contained by the structure, in contrast to the indefinitely upward-stretching void between the Twin Towers that is discussed in chapter 2. Indeed, Koolhaas's mock patented drawing for the structure looks suspiciously like the Twin Towers of the World Trade Center reimagined. Might we even consider this building an entry for the Ground Zero design competition, albeit pasted into a Chinese context?

At only 51 meters high, the CCTV building's gigantism takes a convoluted form, although it offers an impressive 473,000 square meters of office space. The building has no obvious tip and no visible antenna, even though it is a media building whose purpose is to transmit. As the headquarters of Chinese state media—the name "CCTV" strangely mimics that of closed-circuit television surveillance cameras—the building itself seems to embody a gigantic fractured antenna that constantly leaks invisible, intangible

CCTV building in context. © OMA/Jim Gourley.

information. The gigantism of this image is emphasized in Koolhaas's juxtaposition of a picture of the earth with the catchy neologism "Kool China" in his upbeat appraisal of the project in the "Beijing Manifesto." The megalomaniac wordplay—*cool* becomes *Kool* to link with his own name—also attains a quality of transmission. It is as if Koolhaas is projecting a coolness not just onto China, the world's most populous country, but from there to the rest of the world. This gives us an idea of the latent gigantism at play in this project and its antenna ecology: it is a gigantism that takes to an extreme the latent, convoluted, and watchful qualities of the towers we have discussed so far. It showcases a potentially ubiquitous, earth-spanning gigantism that encompasses the capitalist democracies of the Western world as well as their perceived *other*.

Bearing in mind that Koolhaas's move to China might have been personally motivated, it is notable that the "Beijing Manifesto" starts with a world-map-style diagram, with a thick arrow pointing from west to east. Koolhaas states that more tall towers have been built in recent years in Asia than in the United States, arguing that this part of the world has overtaken the West when it comes to building tall. As Marxist geographer David Harvey has argued, the development of towering cities in China became even

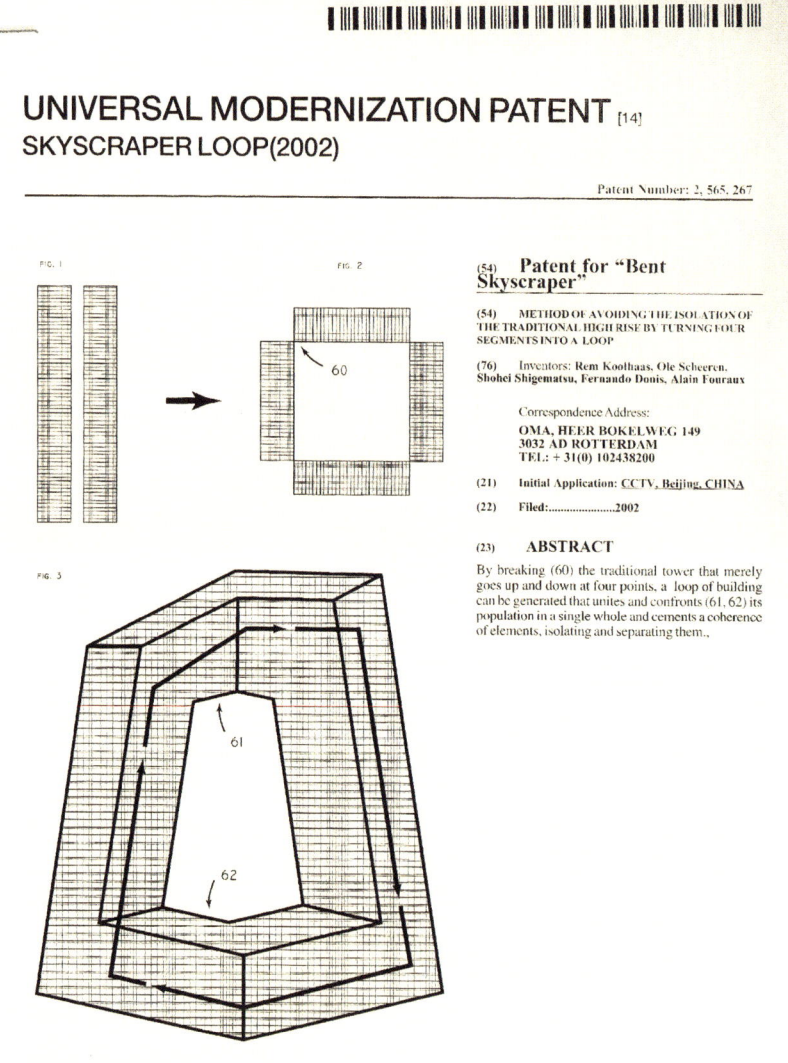

Skyscraper Loop patent. © OMA.

more explosive in the wake of the financial crisis of 2008. Harvey observes that the financial-sector crisis of 2008, which started as a specifically American crisis tied to a specific set of risky subprime mortgages for private buyers of real estate, quickly became a self-perpetuating phenomenon—a traveling crisis with different effects around the globe. In China, Harvey argues, the decrease in jobs following this crisis was met by the government's expansion of the building sector, leading to unfathomable statistical gigantism: in less than three years, between 2011 and 2013, more concrete was used in China than was poured in the United States throughout the entire twentieth century.[12] In 1994 Koolhaas proposed the term *the generic city* to account for, in the first instance, American urban developments of connected, wired, and globally accessible and recognizable urban structures in late capitalist economies in the spirit of the "global village." These are cities that have left behind particular historical, architectural hallmarks in favor of similar, that is, generic, forms and structures—a rhetoric tied to linear and semantic forms of gigantism.[13] However, the gigantic sameness of the Chinese developments of vast areas of tower block cities, despite their sameness and apparently *generic* appearance, is something else when seen as a response to the financial crisis. If these cities are one aspect of a crisis that travels across continents and political cultures, even in their physical manifestation, the recent urban developments in China can be said to culturally linger as a much more perpetual form of crisis, a latent form of gigantism. Seen as one phenomenon, they are unfathomably large, although highly concrete and present when experienced from the ground. This form of development and the problems we have in understanding its impact has to do with what Morton calls the nonlocality of the hyperobject, which can never be expressed or experienced as a unity. It is also related to our argument that latent gigantism is akin to a stowaway (in Gumbrecht and Runia's terminology) in the way it compels us to negotiate a slippery ontological borderline that never quite manifests itself.

In China, the CCTV building has allegedly been given anthropomorphizing nicknames such as "big shorts,"[14] and Koolhaas has felt compelled to counterargue publicly against comments that the building was modeled on a "pornographic position of a woman on her hands and knees."[15] As with Calatrava's Turning Torso, discussed in the prologue to chapter 3, an anthropomorphic imaginary is clearly stirred by this building. It seemingly refers to a human of very great size—a giant, perhaps, like the savage and monstrous mythological figures to which the root of the word *gigantism* points. The particular kinds of monster we envision at a particular time say something about the fears and anxieties of that cultural context.[16] Monstrous figures such as giants often are spun into questions about center and periphery and also point to hidden, unknown forces. In "A Cyborg Manifesto" Haraway notes: "Monsters have always defined the limits of community in Western imaginations. The Centaurs and Amazons of ancient Greece established the limits of the centred polis of the Greek male human by their disruption of marriage and boundary pollutions of the warrior with animality and woman. . . . Cyborg monsters in feminist science fiction define quite different political possibilities and limits from those proposed by the mundane fiction of Man and Woman."[17]

For Haraway, a monstrous figure such as a cyborg can point to a place that offers alternative possibilities for action. With regard to the CCTV building, it is notable that its colloquial reception genders it as female as opposed to the phallic imaginary often tied to the erect skyscraper. Yet the subversive potential of the monstrous that Haraway hopes for somehow eludes the CCTV building, whose monstrosity resides the realm of the latent, which renders its gigantism simultaneously "too much" and "too little." Both Koolhaas and Gumbrecht are implicated in the latency we are after here insofar as they both articulate a particular combination of grandeur and flattening that is discernible in the CCTV building as well as in Gumbrecht's conceptualization of latency as indicative of the postwar period in the Western context.

Furthermore, as architectural critic William Drenttel has argued, Koolhaas's efforts to argue that his building project in China is not just relevant or timely but almost a moral imperative are troublesome in light of the larger crises facing the discipline:

> In the end, all the political discourse and self-serving manifestos mean little. We are left to judge this building as a piece of architecture built in 2007, in a climate of growing awareness of sustainability. Building a project of this scale with so much extra steel to support an aesthetic expression seems like a missed opportunity, if not something completely bordering on civic negligence, especially in China, one of the countries which necessarily must embrace sustainability soon. Imagine if Koolhaas had used this opportunity to build the lightest, most green building in the world? Imagine if he had marshalled all of his rhetorical verve and diplomatic savvy to argue for the critical importance of such architecture? Instead of responding to fortune cookies, Rem Koolhaas could have changed the world.[18]

Like the architecture we discuss under the heading of the new metropolitan mainstream in chapter 3, in his own way Koolhaas has contributed to a turn toward the prevailing condition of architecture and urbanism at the beginning of the twenty-first century, which he addresses under a number of headings (such as dirty realism) and by exploring the potential for architectural beauty in what he calls *junkspace*.[19] Koolhaas's goal is to accept the complexity and lack of centralized control of urban architecture and development in the postindustrial global city.[20] In his writings, we find a call to embrace the loss of identity inherent in globalization as linked to the indefinably large or extra-large and to dispense with any attempt to regain a small-scale relationship with our given surroundings. This is itself a form of gigantism that, like Koolhaas's writings and practice, is a hybrid of linear gigantism (the expansive stardom of the starchitect making big buildings all over the globe) and semantic gigantism (Koolhaas's notorious use of contradictions, his break with the authenticity of formalism and the craft of modernism, and his move instead toward an urbanism on an extra-large scale that is open and networked).

But Koolhaas's playfully and megalomaniacally twisted response to the loss of the Twin Towers, displaced onto the Chinese context, implies a form of gigantism that cannot be broached in dualisms between linear and semantic, closed or open-ended, West and East, or by a narrative where a feeling of personal rejection can be offset by building a large expressive building. Indeed, it implies a form of gigantism that is latent. In the essay "Kill the Skyscraper," Koolhaas hints at this latent gigantism when he writes: "The skyscraper has become less interesting in inverse proportion to its success. It has not been refined but corrupted . . . , negated by repetitive banality."[21] Negation by repetitive banality is at the heart of latent gigantism, and the CCTV building does not escape it. Indeed, when Koolhaas in 2013 received the Skyscraper of the Year award for this building from the Council for Tall Buildings and Urban Habitat, he commented in his usually ironic style: "The fact that I am standing on this stage now, in this position, meant that my declaration of war went completely unnoted, and that my campaign was completely unsuccessful."[22]

Like many of the thinkers and architects we have engaged with in this book, Koolhaas expresses a set of contradictory feelings about tall buildings: they blatantly embody top-down power relations, megalomania, unsustainability, inequality, surveillance, and even paranoia, but they can also lend themselves as vehicles for utopian design strategies and extravagant architectural form experiments. Latent gigantism suggests an explanation for this unresolved uncomfortableness that clings to the cruel optimism of the vertical gigantism of the skyscraper. In a moment, we will take this idea of uncomfortableness back to Ground Zero, but first we explore its less vertical and more horizontal implications in relation to digital tech giants.

Telling Tales of Tech Giants

The architecture of a heightened sense of security at the Ground Zero site is articulated not only in the vocabulary of the spectacle (such as guards and security cameras) but also through a more intrinsic storytelling about the intricate dependencies between the architectural and digital infrastructures that permeate the site. It is an articulation of what we (with reference to surveillance studies scholar David Lyon) can identify as a "culture of surveillance,"[23] and that we can specify (with Wendy Hui Kyong Chun) as a culture of safety that feeds on crisis: "In such a society, each crisis is the motor and the end of control systems; each initially singular emergency is carefully saved, analyzed and codified."[24]

A heightened awareness of risk and security and continual efforts to predict and preempt may turn out to be self-perpetuating, furthering a conception of the world as comprised of complex networks that are fundamentally unstable and in flux. We are dealing here with a highly technologically mediated situation that is concretely embodied in particular practices and through services offered by different companies. This situation needs to be understood as an underlying infrastructure and not only through the smooth spectacles offered, for example, in the media culture we have explored in and around One World Trade Center. These practices and services are a playing field dominated by giants of many sorts, and as Michael Tavel Clarke has noted, there was already at the beginning of the twentieth century a parallel between the critiques of skyscrapers and of big businesses.[25] One of the most marked today is the American company Amazon.

Amazon was named after the largest river in the world, not the tribe of women warriors in Greek mythology. The company's founder, Jeff Bezos (b. 1964), has said that he leafed through the dictionary to find a word beginning with the letter A, and when he came across the word *Amazon*, it seemed fitting because the Amazon "is not only the largest river in the world, it's many times larger than the next biggest river. It blows all other rivers away."[26] The gigantism at work here is about linearity (listed first in the alphabet and quantitatively so much larger than the second-largest that it has the power to "blow away" everything else) and simultaneously also about growing by spilling over into ever-larger areas of commerce. What started in 1994 as an online bookstore is today not just a massive e-commerce marketplace and media platform but also a cloud computing platform that in the US has a market share significantly larger than the second-biggest provider. Indeed, the cloud computing platform has made the biggest contribution to the company's massive rise in revenue in the last couple of years.[27]

The brand name of another American tech giant, Apple, was allegedly likewise chosen partly for its alphabetical priority. According to Apple's cofounder, Steve Jobs (1955–2011), "I was on one of my fruitarian diets. . . . I had just come back from the apple farm. It sounded fun, spirited, and not intimidating. Apple took the edge off the word 'computer.' Plus, it would get us ahead of Atari in the phonebook."[28]

This "fun, spirited, and not intimidating" agenda can be paired with the new metropolitan mainstream in architecture and the corollary flattening of digital culture, a coupling discussed in chapter 3. It has a ubiquity that makes its object so familiar that it becomes almost invisible. With its strong sense of nonlocality, the gigantism of the tech giants bears the hallmarks of the hyperobject. Moreover, global recognizability is at the heart of many tech giants' founding myths, including Amazon and Apple, and point at latent gigantism.

The storytelling behind the Chinese multinational conglomerate Alibaba makes this latent gigantism even more apparent. As its founder, Ma Yun (b. 1964), often referred to under his English name Jack Ma, said in an interview in 2006:

> One day I was in San Francisco in a coffee shop, and I was thinking Alibaba is a good name. And then a waitress came, and I said do you know about Alibaba? And she said yes. I said what do you know about Alibaba, and she said "Open Sesame." And I said yes, this is the name! Then I went onto the street and found 30 people and asked them, "Do you know Alibaba"? People from India, people from Germany, people from Tokyo and China. . . . They all knew about Alibaba. Alibaba—open sesame. Alibaba—40 thieves. Alibaba is not a thief. Alibaba is a kind, smart business person, and he helped the village. So . . . easy to spell, and global know. Alibaba opens sesame for small- to medium-sized companies. We also registered the name AliMama, in case someone wants to marry us![29]

For Ma Yun, Alibaba is a well-chosen name because everyone knows it. It opens doors by being recognizable, seemingly belonging to everyone across cultures, and that is what paves the way for its expansion.

world_record_egg. Creative Commons.

Another example of latent gigantism in digital culture is the world_record_egg. In January 2019, it became the most liked image on Instagram, breaking the previous record of 18 million likes (held by American reality TV star Kylie Jenner with 128 million followers) in under ten days with a generic photo of an egg. When the *New York Times* revealed its creators as three twenty-nine year-old advertising creatives in London, it asked, "Why an egg?" and got the answer "An egg has no gender, race or religion. An egg is an egg, it's universal."[30] The world egg is in fact a motif in the creation myths of many cultures across the globe, representing a containment from which the universe emerges. Yet in its Instagram instantiation, it has become a world record egg, adding the inclination to be first to the holistic imagery. A key reason that the egg succeeded was no doubt the explicitly stated goal of overtaking the existing record. The monumentality of the world_record_egg blows the benign and mundane up to a grand scale. It marks latent gigantism in the way that it does not address inequalities but instead gulps them down by being so large that there should be room for everyone (including any number of stowaways).

A not dissimilar logic marks the rhetoric of big data, which tries to circumvent the risk of bias and overfitting by including everything: n = all. In machine learning algorithms that are trained on these giant databases, however, we see that biases are not done away with but rather reinforced.[31] As Ma Yun, tongue in cheek, notes in his comment on Alimama, Alibaba is a gendered name, leaving us to ponder the potential of its female counterpoint. In November 2007, the Alibaba Group launched www.alimama.com, an online marketing platform that relies on big data. Alimama tracks users and connects the data gathered by the multitude of digital platforms offered by the Alibaba Group (which include demographic attributes, consumption data, physical locations, browsing behaviors, payment methods, and social data) to a customer base that includes practically everyone who uses an electronic device in China, thereby helping advertisers to reach their target groups.[32] In 2018, Alimama reportedly brought in 60 percent of Alibaba's revenue.[33] If Alibaba is "a kind, smart business person" who helps the global village, the mother figure is sitting on a gold mine of knowledge about that village. Resemblances (as in "people like you") and differentials (such as race, gender, ability, or political leanings) come together and are worth a fortune.

Today's tales of tech giants, snippets of which we have recounted here, are ripe with fairytale and cosmological imagery. Like the towers that reach for the sky, the founding myths of these tech giants create links among nature, technology, and the cloud on which both the Alibaba Group and Amazon base much of their operation. They by no means attempt to abandon linear or semantic forms of gigantism, but the most pervasive form of gigantism seems to be the one that has become so latent that we can no longer visualize it, which makes it all the more uncomfortable.

The digital cloud, in this sense, permeates daily life. It is a gigantic industry used by billions of users and accounting for 3 percent of global energy consumption. Despite the metaphor's airy connotations, the digital cloud is not an actual place "up there" in the sky, hovering above us like a meteorological cloud. The digital cloud is concretely embodied in data centers and servers that work and use energy, and because they generate a lot of heat, they need cooling and are therefore intrinsically tied to local climatic conditions and architectural infrastructures.

Moreover, many of those millions of users help to produce what flows in the cloud, making them "prosumers" who simultaneously produce and consume.[34] A notable example is the crowdsourcing marketplace Amazon Mechanical Turk, which solicits users all over the world to help with massive time-consuming tasks such as data validation or content moderation, which are broken into microtasks that can be assigned to people across the internet. Examples include images in which we have to look for objects such as cars or street signs to prove to a website that we are not robots, or the language-learning app Duolingo's initial programming as a way into computerized translation.[35]

So who is helping who? The idea of pitching "the people" against the media or the state, or even pitching humans against technology as was implicit in Koolhaas's narrative, makes much less sense in this context. Although the cloud is largely out of sight, it corresponds to actual sites, people, and structures, and its workings and meanings cannot be separated from local embodying conditions on the ground, which are differentiated, sticky, and fundamentally unequal. If the latency at work here is indeed comparable to that of the stowaway, it is an intricate interplay between vertical and horizontal gigantism across architecture and digital culture that conditions the digital cloud. Cloud services are important players in the antenna ecology, and as a digital infrastructure, the cloud is part architecture and part digital. It is widely distributed and architecturally embedded in servers and cables, and although it professes a neutral

Into the Ground

sameness, it is not the same for everyone but conditions in differentiated manners.

We can store our memories in the cloud. It helps us "share" with others and with our future selves, performing what geographer Louise Amoore calls an "archiving of the future."[36] It is embodied in physical systems, places, and players, like the tech giants we have talked about here, which

Internet Machine 10. © Timo Arnall.

Internet Machine 01. © Timo Arnall.

are as involved in physical places, politics, people, and contestations as everyone else but are doing so on a scale of such gigantic dimensions that their founders would never be able to personally engage with the people and places that form the companies' physical and human makeup. Instead, they and the brands they have created take on superhuman qualities, not unlike the starchitects we have encountered in this book.

In 2017, for example, when Amazon embarked on a mission to establish new additional headquarters in America outside Seattle, the usual relationship between public and private became if not reversed, then certainly reconfigured into a new constellation. It is as if cities bid for Amazon, offering excellent locations and tax breaks in return for the promise of jobs. In the case of New York City, Amazon first accepted this partnership but then decided to quit the deal in light of mounting contestation from the public: "'It was the perfect storm,' Rebecca Kolins Givan, an associate professor in the Rutgers School of Management and Labor Relations, said of Amazon's sudden departure. 'Opposition to Amazon has been brewing nationwide and globally as people realize how much of the retail market they control and the poor quality of jobs they are offering. The strange, secret beauty pageant that Amazon asked cities to participate in created a lot of bad feelings.'"[37]

This brings us back once again not just to New York City but also to wired towers. Glass infrastructures—of tall skyscrapers as much as of the fiber-optic cables that tie together digital infrastructures—are at the heart of surveillance culture. Its operation depends on information transmission, but it brings with it a range of uncertainties that can be conceived as risks as well as opportunities. Most of us willingly and abundantly share our most intimate information when we type search words into Google or engage with social media. Although in recent years we have become increasingly knowledgeable about how we thereby make ourselves targets for online marketing and political manipulation, many of us happily take up residence in metaphorical as well as physical glass buildings, convinced that other people have more interesting things to do than observe us or that there is somehow too much going on, too much information, for anyone to notice us. But the amalgamation of behavior patterns on the scale of the gigantic gives those who want it—corporations, politicians, institutions, or people with their own agendas—endless information, if not about us then about users like us, making us vulnerable both individually and collectively.

It might have taken massive leaks such as WikiLeaks, the Snowden revelations, or the Cambridge Analytica scandal to remind us that we have not moved from nothing to hide and nothing to fear to nothing to hide and nothing to see but rather to nothing to hide and everything to see if you have access. This state of affairs is what we see mirrored in the shifting cultural sentiments and anxieties about our digital traces.[38] These anxieties and traces are not metaphors or symbols, however. Their gigantism is as embodied (and their impact on political and resource-related issues is as real) as any concrete architectural artifact encountered in this book.

Big Business beneath Manhattan

> "I hate to stereotype," says architect Fiona Scott. "Male architects are often quite sensitive, artistic people and any suggestion that buildings designed by women are more curvy, tactile or colourful is wrong. But I don't think there are many women who think, 'Oh, my ideal project would be a massive tower.'"[39]

This book has moved from one tower to the next, towers conceived and carried out mostly by men. It tells a tale of them—a highly selective tale, focusing on a particular Western trajectory, from the Eiffel Tower in Paris to One World Trade Center in New York. But like a tornado moving across the ground, this tale has whirled up a host of issues of inclusion and accessibility—for example, about marginalized groups and in relation to gender, race, and disability or issues concerning ecology and the environment. In this way, it has raised questions about the phallocentric, Eurocentric, and gendered connotations implicit in the anthropocentric inclinations of these structures and the narratives that support them. We are dealing here with biases that create conflicting and uncomfortable interpretations. As the dust settles, we now see different forms of gigantism at work across the board—not just in the wired towers discussed in previous chapters but also in more recent places that widen but also challenge the Western ownership of gigantism, such as Koolhaas's CCTV building in Beijing, and in tech giants' narratives about the fabric of which they are made.

As we have suggested, the notion of the Anthropocene—a way of understanding the relatively new idea that human culture has impacted the earth's ecosystem and climate in fundamental and irreversible ways—can thus be seen as one of contemporary culture's most loaded gigantisms, with an impact on architectural culture as well as on our understanding of human culture more generally. Let us therefore now move back to New York City and approach this discussion in view of latent gigantism to see if it can help us navigate architecture's response to climate change—exemplified in a large resilience project around Manhattan conceived by the architectural firm BIG and in landscape architect James Corner's Field Operations' project at the former waste site called Fresh Kills Landfill.

In chapter 4, we discuss Hurricane Sandy, which severely damaged New York in 2012. According to climate modeling predictions, this kind of storm will not be written into the history books of the future as a stand-alone event. The city is currently preparing itself for further extreme weather events. One of these initiatives is a massive barrier along the southern edge of Manhattan called the BIG U, the realization of which is in progress. According to the New York City Panel on Climate Change, sea levels around the city will increase by between eight and thirty inches by the 2050s and by between fifteen and seventy-five inches by the end of the century. The BIG U is designed to protect Lower Manhattan from floodwater, storms, and other climate change–related extreme weather events. It

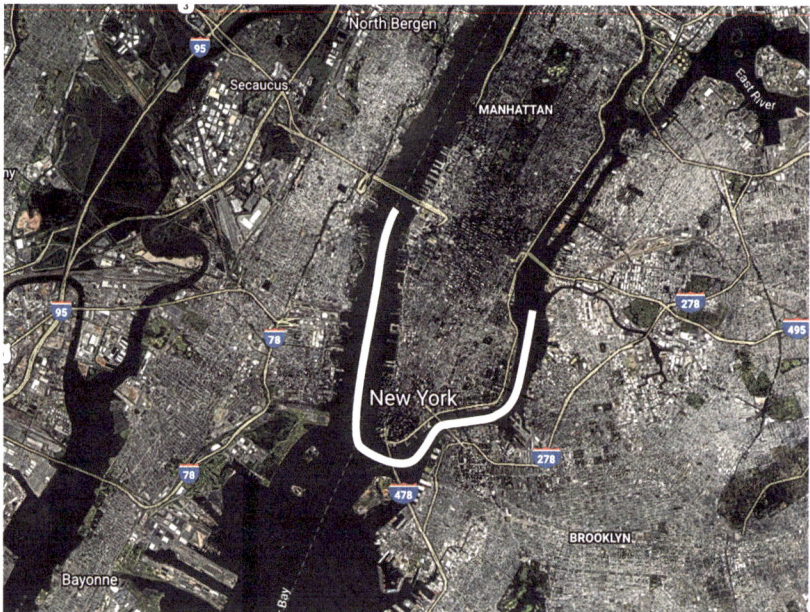

Manhattan U. © Kristen Van Haeren, 2019.

envisions ten continuous miles of protection beginning at West 57th Street, going down to the Battery, and returning back up to East 42nd Street, broken up into three segments: East River Park, Two Bridges and Chinatown, and Brooklyn Bridge to the Battery.[40] The design seemingly has it all: its response to the crisis brought on by climate change will also create a better, more livable, beautiful, safe, and just city. Bjarke Ingels (b. 1974) once worked at OMA for Koolhaas, and like Koolhaas he is a master of striking conceptual design. The BIG U is conceived as a protective cape that folds around the city, creating a form of containment, locking in the city and its grid to keep the water from seeping in. The design integrates flood protection with social and community life by creating a park area with waterfront access and new educational facilities, featuring a reverse aquarium where visitors can observe tidal variations and sea level rises.[41]

Thus, the project showcases the intermingling of security and entertainment that we also saw in the One World Observatory. However, critical voices have been raised. One of these is the New York City Environmental Justice Alliance, a group of organizations that advocate primarily for low-income communities of color. This group has argued that the city lacks

Flooding at Ground Zero. Hurricane Sandy 2012. © John Minchillo / AP / Ritzau Scanpix.

similar commitment levels across other vulnerable waterfront communities to the East Side Coastal Resilience Project of which the BIG U is part.[42] Also, when BIG projects the final World Trade Center building, Two World Trade Center, which will rise up beside One World Trade Center, Ingels seemingly omits any reference to the complex resilience questions involved in building into the sky. He claims this tower will "complete the World Trade Center and finally restore the majestic skyline of Manhattan and unite the streetscapes of TriBeCa with the towers downtown" by stacking Tribeca's characteristic blocks, "forming parks and plazas in the sky" to reach a staggering 1,340 feet, almost the height of the original Twin Towers. According to Ingels, when "horizontal meets vertical, diversity becomes unity."[43] This image of a stacked city reaching into the sky resembles huge tower complexes such as Dubai's Burj Khalifa, the world's tallest tower, which is often referred to as a city in the sky but seems as impenetrable as any walled medieval city. As Joanna Zylinska has aptly noted, the discourse of walls and barriers that work to keep out or contain is threatened by what spills and overflows:

> The mechanism of encystment is regularly mobilized by various sections of the human population in an attempt to ward off all kinds of threats, including the existential threat to our very existence as a species, and to the continued survival of our terrestrial abode. The "cybernetic-sociotechnical shell," which is emerging today in the form of walls, barriers, bans, and exits, is being fueled precisely by discourses of excess, mainly the excess of human bodies—and its organic and nonorganic products. And thus the progressive politics of degrowth on the planetary scale in the face of the Anthropocene finds, perhaps too easily, its ugly twin in the localized discourses of *information and matter overload*: cyberterrorism, multiculturalism, immigration flood, the refugee crisis.[44]

Not only tall towers but also resilience projects such as the BIG U can be seen as parts of larger discourses of containment, overflow, and pertaining dangers. *Resilience* in this context is a temporal term, projecting solidity into the future and thereby granting a sense of agency to the city, the built environment, and the ecological context. Rather like terms such as *smart city*, *resilience* embeds a techno-optimistic, action-oriented temporality where climate change can be managed and dealt with through technological inventions.

Yet when it comes to resilience metaphors, agency is often projected onto agents other than humans, which means that very different outcomes may also be envisioned. The aim may be to create a more livable context for (some) humans, but offering the built environment and ecological players an agency of their own requires the embrace of the very uncertainty from which resilience thinking wishes to turn away. We experience the climate crisis locally whenever an extreme weather event hits the ground on which we stand, yet as a ubiquitous, earth-spanning phenomenon, the gigantism that climate change worryingly marks is a form of latent gigantism, not least because it is not experienced in the same way by everyone. Even if Western industrial culture's impact on the climate potentially affects everyone and everything on the planet, climate change as such can never be seen, felt, experienced, or even noticed as a unified phenomenon by everyone and everything.

When it comes to resilience and climate-adaptation projects such as the BIG U, this situation can be grasped as a form of entrapment. Climate change becomes a necessary driver of change that continues to benefit some at the expense of others and in this gesture justifies the prolonged temporality of resilience thinking itself. Rather than reveal or resolve this form of crisis—and precisely because of the constituents of gigantic latency—we need to approach "it" through the sticky resistances and paradoxes of projects such as the BIG U.

Another project that taps into these questions of long temporal horizon and gives a new perspective on our discussion in chapter 2 of the continued remanence of the Twin Towers is the Freshkills Park project on Staten Island. Here, however, the ethical implications are more physically available from the start, thanks to the connection with the Ground Zero site we know so well. Fresh Kills Landfill was once the world's largest landfill site. After receiving household garbage from Manhattan for more than half a century, it was scheduled to close at the end of 2001. But before it did, it received about a third of the debris from the World Trade Center—around 1.6 million tons of material. In this way, Fresh Kills Landfill became both a cemetery and a sorting site, with thousands of specialists working for more than 1.7 million hours to recover the remains of people killed in the attack.[45] Although three hundred additional individuals were identified from these remains, about a third of those who died in the attack were never accounted for, and it is likely that the debris at Fresh Kills still contains fragmentary human remains.

Fresh Kills Landfill with human remains from the 9/11 attack. © Mike Segar / Ritzau Scanpix.

At present, the site is being reprojected into an urban park by Field Operations, the practice of well-known landscape urbanist James Corner (b. 1961). This transformation is an ongoing process spanning several decades. The ground filled with waste produced by and from humans will be overgrown with a rich green carpet, enabling a multiplicity of new human, animal, and plant life. But we may ask who can guarantee that, in the *long durée* of the process to transform the site—which is packed with the waste of American culture from the height of the industrial period—into sticky, pristine, and uniform earth, no poisonous fumes or liquids and no more ghosts of 9/11 will leak from the ground? On a planetary scale, that *long durée* will be no more than the blink of the eye, and maybe the lush green carpet is in itself an image affirming that there are many other ethics harbored in this site than those pertaining to humans.

We started this book high above the ground at the tip of the Eiffel Tower and atop the World Trade Center and the One World Observatory. We end it in the earth and at the place where water meets the city ground—places where we have to dig through, digest, absorb, and produce that very earth, like earthworms sensing the world through blind, soft bodies. These nonarthropod invertebrate animals stick with and in the earth in a way that is more akin to Donna Haraway's notion of composting than it is to the

Freshkills Park aerial. © James Corner Field Operations.

antenna ecology of Mark Wigley's exoskeletal insects with transmitters on their foreheads, both of which we engage in the introduction to this book.

Projects such as the BIG U or Freshkills Park bring together all the keywords that have run through this book and that constitute the spine of latent gigantism. Containment, leaks, flattening, commonality, crisis, and entrapment are all at work in these imagined structures. However, unlike the other structures that we have looked at—which are bound by the fact that they are actual, physical constructions and that their primary physical appearance is that they reach into the sky—the BIG U and Freshkills Park are vertical structures and will remain in the fictional realm until they are built, and then their gigantism is of a more lateral nature. Although both projects carry an ethos about city life, commonality, and relationships between humans and nature that is both optimistic and caring, they are no less marked by the paradoxes of gigantism we have traced in relation to the towers in this book. Thus, they can be read as our own time's projection of the future, and time will tell how they will negotiate the monstrosity that comes with gigantism—its dangers and its potentials. Indeed, the monstrous is grandeur's twin, and in gigantism they come together to form the monumental—something that we hope will stand when all else fails. In that sense, latent gigantism represents a broad presencing that makes the monstrous times we envision an inextricable part of the present.

Epilogue: The Bias Cut

Copenhagen, March 2019

I neatly fold baby clothes that my daughter has outgrown. Growth is a strange thing. Clothes that were too big for her only a few months ago now make her look like a giant. Most of the clothes were bought online, and although of different brands they are predominantly "made in China." Some were inherited from Henriette's daughter, who is exactly one year older than mine, to the day—one-year-apart twins. Some of them are clothes I wore myself when I was my daughter's age—garments knitted, sewn, or crocheted by my mother or grandmother, by women's hands more dexterous than mine. They say it is a skill that runs in the family. My great-grandmother, born around the time the Eiffel Tower was erected, supported her five children and bankrupt husband by working as a seamstress. But although I have learned all the techniques, my fingers are intuitively more comfortable scrolling, swiping, tweeting, texting, and snapping than weaving, knitting, or mending.

Despite the current trends for gender-neutral names or for play with gendered fashion markers, baby clothes are still an area where it is hard to escape the normativity of the pink-blue binary. But when baby clothes work best, they dress a body that bears only largely hidden signs of the child's sex, and they allow unrestricted movement. As I let my finger rub against the fabric of a purple playsuit, I cannot help thinking about the fact that the word *bias* has a second meaning, beside the one that prevails in an age of machine learning and artificial intelligence. With regard to textiles, *bias* is used to describe a cut diagonal to the weave of a fabric. The grid structure of horizontal weft woven through vertical warp determines how a fabric moves, but a diagonal cut at a 45-degree angle across that structure will allow the fabric to flow in a different, more fluid, almost leaking manner. The fabric will be more elastic, requiring more technical skill to hold it in place and avoid puckering the seam, but it will also drape itself more softly along the curves of the body.

Fabric is at the same time a vehicle for grand and spectacular fashion statements *and* the material of anonymity and fragility. It may seem ephemeral, and we might assume that it will decompose much sooner than the bones of our skeleton or the concrete pillars of the architecture that surrounds us. Yet this is not necessarily the case. Biodegradability is highly dependent on context and environmental conditions. For instance, natural fibers such as wool and cotton decompose more readily than polyester or nylon, and new garments can be made by recycling old fabrics, allowing something else to emerge. My grandmother, after whom my daughter is named, used to recite the epic poetry of the Finnish *Kalevala*, which describes the birth of the earth from the shards of an egg. As I put the playsuit into the box and close the lid, the words of the poem that I have heard so many times keep reverberating.

—KV

towertotower
Copenhagen

But the eggs and pieces were not
Mixed up with the mud and water
For at once the crumbs grew comely
And the pieces beautiful.

One egg's lower half transformed
And became the earth below,
And its upper half transmuted
And became the sky above;
From the yolk the sun was made,
Light of day to shine upon us;
From the white the moon was formed,
Light of night to gleam above us;
All the colored brighter bits
Rose to be the stars of heaven
And the darker crumbs changed into

Clouds and cloudlets in the sky.

● Liked by **mariafinn_** and **2 others**
towertotower Kalevala #biascut #worldegg #brighterbits #cloud

Notes

Introduction

1. British Standards Institute, "Temporary Works Equipment. Scaffolds. Performance Requirements and General Design," BS EN 12811-1:2003 (2004).

2. Charles Dickens, *A Tale of Two Cities* (London: Penguin, 1989), 35.

3. Martin Heidegger, "The Age of the World Picture" (1938), in *The Question Concerning Technology and Other Essays* (New York: Harper Torchbooks, 1977), 115–154.

4. Anna Lowenhaupt Tsing, *The Mushroom at the End of the World: On the Possibility of Life in Capitalist Ruins* (Princeton: Princeton University Press, 2015), 17–25, 37–43.

5. Daniel R. Headrick, *The Invisible Weapon: Telecommunications and International Politics, 1851–1945* (New York: Oxford University Press, 2012), 123.

6. We have found encouragement and inspiration in Nina Lykke, ed., *Writing Academic Texts Differently: Intersectional Feminist Methodologies and the Playful Art of Writing* (New York: Routledge, 2014), as well as in Donna Haraway, "Situated Knowledges: The Science Question in Feminism and the Privilege of Partial Perspective," *Feminist Studies* 14, no. 3 (1988): 575–599.

7. *Oxford English Dictionary*, s.v. "gigantism."

8. Anna Lowenhaupt Tsing, "On Nonscalability: The Living World Is Not Amenable to Precision-Nested Scales," *Common Knowledge* 18, no. 3 (2012): 523.

9. Rob Nixon, *Slow Violence and the Environmentalism of the Poor* (Cambridge, MA: Harvard University Press, 2013).

10. Margaret Atwood, *Oryx and Crake* (New York: Anchor Books, 2003), 243.

11. Donna Haraway, *Staying with the Trouble: Making Kin in the Chthulucene* (Durham, NC: Duke University Press, 2016), 4.

12. See e.g. Donna Haraway, "Situated Knowledges: The Science Question in Feminism and the Privilege of Partial Perspective," *Feminist Studies* 14, no. 3 (1988): 579.

13. Lauren Berlant, *Cruel Optimism* (Durham, NC: Duke University Press, 2011), 1.

14. Timothy Morton, "From Modernity to the Anthropocene: Ecology and Art in the Age of Asymmetry," *International Social Science Journal* 63, nos. 207–208 (2012): 47; Timothy Morton, *Hyperobjects* (Minneapolis: University of Minnesota Press, 2013).

15. Michael Tavel Clarke and David Wittenberg, "Introduction," in *Scale in Literature and Culture*, ed. Michael Tavel Clarke and David Wittenberg (London: Palgrave Macmillan, 2017), 7.

16. Heidegger, "The Age of the World Picture," 135.

17. For an overview of more technical critiques of the high-rise, see Kheir Al-Kodmany and Mir M. Ali, *The Future of the City: Tall Buildings and Urban Design* (Champaign: University of Illinois, 2012).

18. See, e.g., Wendy Hui Kyong Chun, "Big Data as Drama," *ELH* 83, no. 2 (2016): 366–382.

19. Ruben Baart, "Interview Mark Wigley: 'We Are Living in an Ecology of Antenna,'" *Next Nature Network*, March 17, 2018, https://tinyurl.com/y58559ct.

20. Mark Wigley, *The Human Insect: Antenna Architectures 1887–2017* (Rotterdam: Het Nieuwe Instituut), accessed May 16, 2019, https://tinyurl.com/y3o5gqcj.

21. See, e.g., Denis Jamet, "What Do Internet Metaphors Reveal about the Perception of the Internet?," *Metaphorik.de*, no. 18 (2010): 17–32.

22. See, e.g., Daniela Agostinho, Catherine D'Ignazio, Annie Ring, Nanna Thylstrup, and Kristin Veel, "Uncertain Archives: Approaching the Unknowns, Errors, and Vulnerabilities of Big Data through Cultural Theories of the Archive," *Surveillance and Society* 17, no. 3 (2019).

23. Council on Tall Buildings and Urban Habitat, "100 Tallest Completed Buildings in the World by Height to Architectural Top," Global Tall Building Database of the CTBUH, accessed May 17, 2019, https://tinyurl.com/y5m9jrqt.

24. Sianne Ngai, *Our Aesthetic Categories: Zany, Cute, Interesting* (Cambridge, MA: Harvard University Press, 2012).

25. Susan Stewart, *On Longing: Narratives of the Miniature, the Gigantic, the Souvenir, the Collection* (Durham, NC: Duke University Press, 1993).

26. We have discussed variants of this slide in previous work: Henriette Steiner, "Nature Created? Or, the Gentle Touch of Artificial Snow," *Montreal Architectural Review*, no. 4 (2017): 5–18; Henriette Steiner and Kristin Veel, "Towering Invisibilities: A Cultural Theoretical Reading of the Eiffel Tower and the One World Trade Center," *Qualitative Inquiry* 25, no. 4 (2019): 407–416.

27. Ian Hodder, "The Entanglements of Humans and Things: A Long-Term View," *New Literary History* 45, no. 1 (2014): 25.

28. Sara Ahmed, *The Cultural Politics of Emotion* (Edinburgh: Edinburgh University Press, 2004); Zoë Sofoulis, "Container Technologies," *Hypatia* 15, no. 2 (2000): 188.

29. Ursula Le Guin, "The Carrier Bag Theory," in *The Ecocriticism Reader: Landmarks in Literary Ecology*, ed. Cheryll Glotfelty and Harold Fromm (Athens: University of Georgia Press, 1996).

30. Sofoulis, "Container Technologies," 188.

31. Sofoulis, "Container Technologies," 188.

32. See, e.g., Wendy Hui Kyong Chun and Sarah Friedland, "Habits of Leaking: Of Sluts and Network Cards," *Differences* 26, no. 2 (2015): 1–28; Daniela Agostinho and Nanna Bonde Thylstrup, "If Truth Was a Woman: Leaky Infrastructures and the Gender Politics of Truth Telling," *Ephemera: Theory and Politics in Organization* 19, no. 4 (2019).

33. Sofoulis, "Container Technologies," 192.

34. Ahmed, *The Cultural Politics of Emotion*, 11, 90.

35. See, e.g., Tsing, "On Nonscalability."

36. Michael Tavel Clarke, *These Days of Large Things: The Culture of Size in America 1865–1930* (Ann Arbor: University of Michigan Press, 2009).

37. Joseph Harriss, *The Tallest Tower: Eiffel and the Belle Epoque* (Boston, MA: Houghton Mifflin, 1975), 50.

38. Marshall McLuhan, *Understanding Media: The Extensions of Man* (Berkeley, CA: Gingko Press, 2011).

39. This and related terms have a long history in critiques of modern society, such as of consumer or entertainment culture. In relation to Danish philosopher Søren Kierkegaard's concept of "leveling," see Henriette Steiner, *Emergence of a Modern City* (London: Routledge, 2014).

40. Shannon Mattern, *Code and Clay, Data and Dirt: Five Thousand Years of Urban Media* (Minneapolis: University of Minnesota Press, 2017), xl.

41. Peter Carl, "Convivimus Ergo Sumus," in *Phenomenologies of the City: Histories and Philosophies of Architecture*, ed. Henriette Steiner and Max Sternberg (New York: Routledge, 2015), 1–31.

42. Heidegger, "The Age of the World Picture," 135.

43. Reinhart Koselleck, *Futures Past: On the Semantics of Historical Time* (New York: Columbia University Press, 2004), 255–275.

44. The word has recently received considerable interest in aesthetic and cultural theory. See, e.g., Hans Ulrich Gumbrecht and Florian Klinger, *Latenz: Blinde Passagiere in den Geisteswissenschaften* (Göttingen: Vandenhoeck and Ruprecht, 2011). See also Kristin Veel, "Latency," in *Uncertain Archives: Critical Keywords for Big Data*, ed. Nanna Bonde Thylstrup, Daniela Agostinho, Annie Ring, Catherine D'Ignazio, and Kristin Veel (Cambridge, MA: MIT Press, 2020).

45. Hans Ulrich Gumbrecht, *Our Broad Present: Time and Contemporary Culture* (New York: Columbia University Press, 2014), xiii. Here Gumbrecht draws on Kosselleck's work on the emergence of fundamental modern categories. However, the notion of the broad present resonates across a set of diverging contemporary theoretical positions, among these Peter Osborne, "The Postconceptual Condition: Or, the Cultural Logic of High Capitalism Today," *Radical Philosophy*, no. 184 (2014): 18–27; Geoff Cox and Jacob Lund, *The Contemporary Condition: Introductory Thoughts on Contemporaneity and Contemporary Art* (Berlin: Sternberg Press, 2016); Wolfgang Ernst, *Chronopoetics: The Temporal Being and Operativity of Technological Media* (Lanham, MD: Rowman and Littlefield, 2016).

46. See, for instance, Louise Amoore and Volha Piotukh, *Algorithmic Life: Calculative Devices in the Age of Big Data* (London: Routledge, 2015).

47. Peter Engelke and J. R. McNeill, *The Great Acceleration: An Environmental History of the Anthropocene Since 1945* (Cambridge, MA: Harvard University Press).

48. Paul J. Crutzen, "Geology of Mankind," *Nature, International Journal of Science* 415 (2002): 23.

49. See, e.g., Andreas Malm, *The Progress of This Storm: Nature and Society in a Warming World* (London: Verso, 2017); Nixon, *Slow Violence and the Environmentalism of the Poor*; Kathryn Yussof, *A Billion Black Anthropocenes or None* (Minneapolis: University of Minnesota Press, 2018).

50. N. Katherine Hayles, *Unthought: The Power of the Cognitive Nonconcious* (Chicago: Unversity of Chicago Press, 2017).

Chapter 1

1. Roland Barthes, "The Eiffel Tower," in *Rethinking Architecture: A Reader in Cultural Theory*, ed. Neil Leach (London: Routledge, 1997), 166.

2. Joseph Harriss, *The Tallest Tower: Eiffel and the Belle Epoque* (Boston, MA: Houghton Mifflin, 1975).

3. Daniel R. Headrick, *Invisible Weapon: Telecommunications and International Politics, 1851–1945* (New York: Oxford University Press, 2012).

4. See, e.g., "On the Concept of History," in *Walter Benjamin: Selected Writings 4: 1938–40*, ed. Howard Eiland and Michael W. Jennings (Cambridge, MA: Harvard University Press, 2006), 389–411.

5. Walter Benjamin, *The Arcades Project* (Cambridge, MA: Harvard University Press, 2002).

6. Benjamin, *The Arcades Project*, 885–887.

7. Detlef Mertins, "Walter Benjamin and the Tectonic Unconscious: Using Architecture as an Optical Instrument," *Departmental Papers (Architecture)*, no. 9 (1999): 199.

8. Benjamin, *The Arcades Project*, 887.

9. Benjamin, *The Arcades Project*, 221 (I4, 4). See also Walter Benjamin, *A Berlin Childhood around 1900* (Cambridge, MA: Harvard University Press, 2006).

10. Benjamin, *The Arcades Project*, 163 (F5a, 7). Benjamin quotes from Egon Friedell's 1927–1931 *Kulturgeschichte der Neuzeit* (Munich: C. H. Beck, 2008), 363.

11. See Richard Hamman and Jost Hermand, *Stilkunst um 1900* (Berlin: Akademie Verlag, 1967).

12. See, e.g., Barry Bergdoll, "Introduction," in Lucien Hervé, *The Eiffel Tower* (Princeton: Princeton Architectural Press, 2003).

13. Harriss, *The Tallest Tower*, 14.

14. Benjamin, *The Arcades Project*, 887.

15. Benjamin, *The Arcades Project*, 161 (F4a, 4). Benjamin here quotes from Lucien Dubech and Pierre d'Espezel's *Histoire de Paris* (Paris: Payot, 1926), 461–462.

16. Dalibor Vesely, *Architecture in the Age of Divided Representation: The Question of Creativity in the Shadow of Production* (Cambridge, MA: MIT Press, 2004), 305.

17. Benjamin, *The Arcades Project*, 160–161 (F4a, 2), our emphasis. The quotation at the end of this passage is from Alfred Gotthold Meyer, *Eisenbauten: Ihre Geschichte und Æsthetik* (Esslingen: Paul Neff, 1907), 93.

18. Walter Benjamin, "The Work of Art in the Age of Mechanical Reproduction," in *Illuminations* (New York: Schocken, 2007), 250.

19. Benjamin, *The Arcades Project*, 459 (N1a, 1).

20. Benjamin, *The Arcades Project*, 160–161 (N2, 6).

21. Benjamin, *The Arcades Project*, 463 (N3, 1).

22. Barthes, "The Eiffel Tower," 165.

23. See also our article "Towering Invisibilities," which can be read as a pilot study for this book.

24. Barthes, "The Eiffel Tower," 166.

25. Barthes, "The Eiffel Tower," 166.

26. Barthes, "The Eiffel Tower," 171.

27. Barthes, "The Eiffel Tower," 171.

28. Barthes, "The Eiffel Tower," 166.

29. Barthes, "The Eiffel Tower," 165–166.

30. Donna Haraway, "A Cyborg Manifesto: Science, Technology, and Socialist-Feminism in the Late Twentieth Century," in *Simians, Cyborgs and Women: The Reinvention of Nature* (London: Routledge, 1991), 177.

31. Jean Jullien, "Symbol des Friedens für Paris," *Spiegel Online*, November 15, 2015, https://tinyurl.com/o9fwl8c.

32. Sam Webb, "ISIS Calls for Attacks in Berlin and Brussels to 'Paralyse' Europe in Wake of Brexit Chaos," *The Mirror*, June 24, 2016. However, its status as an ISIS production remains unconfirmed, which speaks to the fundamental ambiguity of such viral pieces. See also Thorsten Botz-Bornstein, "The 'Futurist' Aesthetics of ISIS," *Journal of Aesthetics and Culture* 9, no. 1 (2017): 1–13.

33. Stephanie Glaser, "The Eiffel Tower: Cultural Icon, Cultural Interface," in *Cultural Icons*, ed. Keyan G. Tomaselli and David Scott (Aarhus: Intervention Press, 2009).

34. See Eric Snodgrass, *Executions: Power and Expression in Networked and Computational Media* (Malmö: Malmö University, 2017), 169, for an in-depth discussion of abstraction as a social media strategy.

35. See, e.g., danah boyd, "Social Network Sites as Networked Publics: Affordances, Dynamics, and Implications," in *Networked Self: Identity, Community, and Culture on Social Network Sites*, ed. Zizi Papacharissi (New York: Routledge, 2010), 39–58.

36. Wendy Hui Kyong Chun and Sarah Friedland, "Habits of Leaking: Of Sluts and Network Cards," *Differences* 26, no. 2 (2015): 4.

37. Chun and Friedland, "Habits of Leaking," 5.

38. Shoshana Zuboff, *The Age of Surveillance Capitalism* (New York: Public Affairs, 2019).

39. Chun and Friedland make this argument with regard to young girls who are slut-shamed, but see also, e.g., Lisa Nakamura and Peter A. Chow-White, eds., *Race after the Internet* (London: Routledge, 2011).

40. Wendy Hui Kyong Chun, "Crisis, Crisis, Crisis, or Sovereignty and Networks," *Theory, Culture and Society* 28, no. 6 (2011): 95.

41. Reinhart Koselleck, *Critique and Crisis* (Cambridge, MA: MIT Press, 1988); François Hartog, *Regimes of Historicity: Presentism and Experiences of Time* (New York: Columbia University Press, 2015), xxviii–xv. See also *The Cultural Life of Catastrophes and Crises*, ed. Carsten Meiner and Kristin Veel (Berlin: De Gruyter, 2012).

42. Reinhart Koselleck, *Futures Past: On the Semantics of Historical Time* (New York: Columbia University Press, 2004), 59–63. See also Poul F. Kjaer and Niklas Olsen, *Critical Theories of Crisis in Europe: From Weimar to the Euro* (Lanham, MD: Rowman and Littlefield, 2016).

43. Hartog, *Regimes of Historicity*, xxviii.

44. Jay David Bolter, *The Digital Plenitude: The Decline of Elite Culture and the Rise of New Media* (Cambridge, MA: MIT Press, 2019).

45. Berlant, *Cruel Optimism*.

Chapter 2

1. National Research Council of the National Academies, *The Internet under Crisis Conditions: Learning from September 11* (Washington, DC: National Academies Press, 2003), https://www.nap.edu/read/10569/, 14.

2. National Research Council, *The Internet under Crisis*, 24.

3. Mark Hall and Lucas Mearian, "IT Focus Turns to Disaster Recovery," *IDG*, September 11, 2001, https://tinyurl.com/y3xg77qe. It is worth bearing in mind that the internet looked significantly different in 2001 than it does today.

4. National Research Council, *The Internet under Crisis*, 25.

5. National Research Council, *The Internet under Crisis*, 27.

6. Jan Jack Gieseking, "Size Matters to Lesbians Too: Queer Feminist Interventions into the Scale of Big Data," *Professional Geographer* 70 (2018): 150–156.

7. Hilary Charlesworth and Christine Chinkin, "Sex, Gender and September 11," *American Journal of International Law*, no. 96 (2002): 600–605; Jan Jindy Pettmann, "Feminist International Relations after 9/11," *Brown Journal of World Affairs* 10, no. 2 (2004): 85–96; J. Ann Thickner, "Feminist Perspectives on 9/11," *International Studies Perspectives* 3, no. 4 (2002): 333–350.

8. Geraldine Pratt and Victoria Rosner, eds., *The Global and the Intimate: Feminism in Our Time* (New York: Columbia University Press, 2002).

9. Pettman, "Feminist International Relations," 91.

10. Zoë Sofoulis, "Container Technologies," *Hypatia* 15, no. 2 (2000): 188.

11. James Glanz and Eric Lipton, *City in the Sky: The Rise and Fall of the World Trade Center* (New York: Times Books, 2004), 28.

12. Quoted in Emmanuel Petit, *Irony, or, the Self-Critical Opacity of Postmodern Architecture* (New Haven: Yale University Press, 2013), 4.

13. Neal Bascomb, *Higher: The Historic Race to the Sky and the Making of the City* (New York: Random House, 2003).

14. See, e.g., Carol Willis, *Form Follows Finance: Skyscrapers and Skylines in New York and Chicago* (New York: Princeton Architectural Press, 2005).

15. Eric Darton, *Divided We Stand: A Biography of the World Trade Center* (New York: Basic Books, 2000).

16. David Gordon, *Battery Park City: Politics and Planning on the New York Waterfront* (London: Routledge, 1997). Yet the park is on a welcoming human scale, with visible traces of everyday life—complete with public institutions such as a library, a school, hotels, restaurants, and places of entertainment.

17. Martin Zerlang, "Urban Life as Entertainment," in *The Urban Lifeworld: Formation, Perception, Representation*, ed. Peter Madsen and Richard Plunz (London: Routledge, 2002), 316.

18. Paul Virilio, *Ground Zero* (London: Verso, 2002), 82.

19. Paul Virilio, "Disorientation," in *Architecture Principe: 1966 and 1996*, ed. Paul Virilio and Claude Parent (Besançon: Les Éditions de l'imprimeur, 1997), 7–8.

20. See, e.g., Julian Reid, *Biopolitics of the War on Terror: Life Struggles, Liberal Modernity and the Defence of Logistical Societies* (Manchester: Manchester University Press, 2013), for a discussion of the debate on the interrelations between Foucault and Virilio.

21. We may approach Virilio's critique of the vertical in a vein similar to how Deleuze approaches Foucault's notion of the disciplinary society. Deleuze argues that the disciplinary society of enclosure has been replaced by a society of control that is characterized above all by movement and modulation yet that is no less oppressive. Gilles Deleuze, "Postscript on the Societies of Control," *October*, no. 59 (1992): 3–7. See also Jeanne Haffner, *A View from Above* (Cambridge, MA: MIT Press, 2013).

22. Michel de Certeau, *The Practice of Everyday Life* (Oakland: University of California Press, 1984), 91.

23. Certeau, *The Practice of Everyday Life*, 92.

24. Certeau, *The Practice of Everyday Life*, 92, our emphasis.

Notes to Chapter 2

25. Certeau, *The Practice of Everyday Life*, 93.

26. Donna Haraway, "Situated Knowledges," 581.

27. Haraway, "Situated Knowledges," 584.

28. Ben Highmore has argued that Certeau's writings in general and his work on the World Trade Center in particular establish a set of dualisms that are subsequently deconstructed and relativized in the texts. Ben Highmore, *Everyday Life and Cultural Theory: An Introduction* (London: Routledge, 2002), 154.

29. Haraway, "Situated Knowledges," 584.

30. In *Complexity and Contradiction in Architecture* (New York: Museum of Modern Art, 1966), Robert Venturi—one of the most significant postmodern architects—describes postmodern architecture as characterized by a plethora of meaning: "I speak of a complex and contradictory architecture based on the richness and ambiguity of modern experience, including that experience which is inherent in art. . . . I like elements which are hybrid rather than 'pure,' compromising rather than 'clean.' . . . I prefer 'both-and' to 'either-or,' black and white, and sometimes gray, to black or white" (16).

31. See Joseph Rykwert, *The Seduction of Place* (Oxford: Oxford University Press, 2000), 247–266. See also the 1972 promotional video for the World Trade Center: *World Trade Center Construction Promo* (1968–1972), accessed May 16, 2019, https://tinyurl.com/y67wjbo6.

32. See, e.g., Angus Kress Gillespie, *Twin Towers: The Life of New York City's World Trade Center* (New Brunswick, NJ: Rutgers University Press, 1999).

33. Sofoulis, "Container Technologies," 188.

34. In a letter, the buildings' architect, Minoru Yamasaki, denounced modernist architecture's foible for glass and its later corporate variants, saying that "the days of the all-glass buildings are finished." Yamasaki defends his slim, high windows, which invoke neo-Gothic motifs as opposed to modernism's panoramic windows and curtain walls, thereby equating his criticism of glass architecture with a critique of architectural modernism. See also Eve Blau, "Transparency and the Irreconcilable Contradictions of Modernity," *Praxis: Journal of Writing + Building*, no. 9 (2007): 50–59.

35. Minoru Yamasaki, letter exhibited in the Skyscraper Museum, New York City.

36. Petit, *Irony*, 5.

37. Eric Frampton, *Modern Architecture and the Critical Present* (London: Architectural Design, 1982).

38. Petit, *Irony*, 5.

39. Patrick M. Bray, "Aesthetics in the Shadow of No Towers: Reading Virilio in the Twenty-First Century," *Yale French Studies*, no. 114 (2008): 4–17.

40. Ada Louise Huxtable, "Who's Afraid of the Big Bad Buildings?," *New York Times*, May 29, 1966.

41. As media theorist Lee Rodney remarks: "Tall buildings and web-cameras are partners in the architecture of transparency; the drive for an expanded horizon links both the aspirations of the glass tower and the surveillance image." Lee Rodney, "Real Time, Catastrophe, Spectacle," in *The Spectacle of the Real*, ed. Geoff King (Portland: Intellect, 2005), 39.

42. Reid, *Biopolitics*, 95.

43. Slavoj Žižek, "Welcome to the Desert of the Real!," in *Dissent from the Homeland: Essays after September 11*, ed. Stanley Hauerwas and Frank Lentricchia (Durham, NC: Duke University Press, 2003), 132–133.

44. Reid, *Biopolitics*, 92.

45. Jean Baudrillard, *The Spirit of Terrorism* (New York: Verso Books, 2012), 5–6.

46. In a biopolitical optic, Reid notes: "Rather than thinking of the event of 9/11 as the first act in a new era of a war between liberal regimes and their outside, we can better think of it as a culminating act in an older war against these architectural techniques of liberal regimes for the control of the life of their populations." Reid, *Biopolitics*, 86.

47. Karlheinz Stockhausen, cited in "The Demolition of the World Trade Center," Radical Art, accessed May 16, 2019, https://tinyurl.com/y3r2yszq.

48. Darton, *Divided We Stand*, 152.

49. Regine Prange, "The Crystalline: Gothic Visions of Architecture: Lyonel Feninger and Karl Friedrich Schinkel," in *The Romantic Spirit in German Art 1790–1990*, ed. Keith Hartley, Henry Meyric Hughes, William Vaughan, and Peter-Klaus Schuster (Stuttgart: Oktagon Verlag, 1994), 155–163.

50. Jean Baudrillard and Jean Nouvel, *The Singular Objects of Architecture* (Minneapolis: University of Minnesota Press, 2002), 38.

51. Haraway, "A Cyborg Manifesto," 67.

Chapter 3

1. See also the discussions in Norman K. Denzin and Yvonna S. Lincoln, eds., *9/11 in American Culture* (Lanham, MD: AltaMira Press, 2003).

2. See, e.g., Diana Gonçalves, *9/11: Culture, Catastrophe and the Critique of Singularity* (Berlin: De Gruyter, 2016).

3. Elizabeth Greenspan sums up these critiques in *Battle for Ground Zero: Inside the Political Struggle to Rebuild the World Trade Center* (London: Palgrave Macmillan, 2013).

4. Lauren Kogod and Michael Osman, "Girding the Grid: Abstraction and Figuration at Ground Zero," *Grey Room*, no. 13 (2003): 15.

5. Charles Jencks, *The New Paradigm in Architecture* (New Haven, CT: Yale University Press, 2002).

6. David V. Dunlap, "Memorial Pools Will Not Quite Fill Twin Footprints," *New York Times*, December 15, 2005.

7. Council on Tall Buildings and Urban Habitat, "One World Trade Center," Global Tall Building Database of the CTBUH, accessed May 18, 2019, https://tinyurl.com/y5satl75.

8. Christian Schmid, "Henri Lefebvre, the Right to the City and the New Metropolitan Mainstream," in *Cities for People, Not for Profit: Critical Urban Theory and the Right to the City*, ed. Neil Brenner, Peter Marcuse, and Margit Mayer (London: Routledge, 2012), 42–62.

9. Schmid, "Henri Lefebvre," 54.

10. Schmid, "Henri Lefebvre," 55.

11. Maroš Krivý and Leonard Ma, "The Limits of the Livable City: From Homo Sapiens to Homo Cappuccino," *Avery Review*, no. 30 (2018): 1–10, https://tinyurl.com/y3s337ef.

12. Henriette Steiner and Natalie Gulsrud, "When Urban Greening Becomes an Accumulation Strategy: Exploring the Ecological, Social, and Economic Calculus of the High Line," *Journal of Landscape Architecture*, no. 3 (2019): 38–43.

13. Jean Baudrillard, "L'Ésprit du terrorisme," *Le Monde*, November 2, 2001.

14. Peter Weibel, "Pleasure and the Panoptic Principle," in *CTRL [SPACE]: Rhetorics of Surveillance from Bentham to Big Brother*, ed. Thomas Y. Levin, Ursula Frohne, and Peter Weibels (Cambridge, MA: MIT Press, 2002), 206–223.

15. Michel Foucault, *Discipline and Punish: The Birth of the Modern Prison* (New York: Vintage, 1995), 195–209.

16. Alan Yuhas, "One World Trade Center Elevators Offer 500-Year History Ride—in 47 Seconds," *The Guardian*, April 20, 2015.

17. Ed Pilkington and Tom Pietrasik, "One World Trade Observation Deck Opens to the Public: Video Preview," *The Guardian*, May 28, 2015.

18. See, e.g., Natasha Dow Schüll, *Addiction by Design: Machine Gambling in Las Vegas* (Princeton, NJ: Princeton University Press, 2012).

19. Mario Carpo, *The Alphabet and the Algorithm* (Cambridge, MA: MIT Press, 2011), 106–107.

20. See, e.g., Peter Marcuse, "The Ground Zero Architectural Competition: Designing without a Plan," *PN: Planners Network*, January 22, 2002, https://tinyurl.com/y6msm2ts.

21. See Marrikka Trotter, "Get Fit: Morphosis's New Academic Building for the Cooper Union," *Harvard Design Magazine* 31 (2009).

22. Gary Hustwit, ed., *Helvetica/Objectified/Urbanized: The Complete Interviews* (London: Versions, 2015).

23. Hustwit, *Helvetica/Objectified/Urbanized*.

24. Carpo, *The Alphabet*.

25. Jean-François Lyotard, *The Postmodern Condition: A Report on Knowledge* (Minneapolis: University of Minnesota Press, 1984).

26. Francis Fukuyama, *The End of History and the Last Man* (New York: Free Press, 1992). See also Christopher Hughes, *Liberal Democracy as the End of History* (London: Routledge, 2012).

27. Mikkel Bolt Rasmussen, *Hegel after Occupy* (Berlin: Sternberg Press, 2018), 11.

28. Saidiya Hartman, *Wayward Lives, Beautiful Experiments: Intimate Histories of Social Upheaval* (New York: Norton, 2019).

29. Hartman, *Wayward Lives*, xv.

30. See, for example, Bruno Latour, *Facing Gaia: Eight Lectures on the New Climatic Regime* (Cambridge: Polity Press, 2017). See also Isabelle Stengers, *In Catastrophic Times: Resisting the Coming Barbarism* (Ann Arbor: Open Humanities Press/meson press, 2015); Danielle Sands, "Gaia, Gender, and Sovereignty in the Anthropocene," *philoSOPHIA* 5, no. 2 (2015).

31. Zoe Todd, "An Indigenous Feminist's Take on the Ontological Turn: 'Ontology' Is Just Another Word for Colonialism," *Journal of Historical Sociology*, no. 29 (2016): 8.

32. Glen Sean Coulthard, *Red Skin, White Masks: Rejecting the Colonial Politics of Recognition* (Minneapolis: University of Minnesota Press, 2014), 176.

Notes to Chapter 4

33. See for instance, Claire Colebrook, "What Is the Anthropo-Political?," in *Twilight of the Anthropocene Idols*, ed. Tom Cohen, Claire Colebrook, and J. Hillis Miller (London: Open Humanities Press, 2016), 91.

34. Joanna Zylinska, *The End of Man: A Feminist Counterapocalypse* (Minneapolis: University of Minnesota, 2018), 59.

35. As noted by, among others, Sara Ahmed with regard to cartographic space. See Sara Ahmed, *Queer Phenomenology: Orientations, Objects, Others* (London: Duke University Press, 2006), 113.

Chapter 4

1. Patrick Howell O'Neill, "Leaked NSA Tools, Now Infecting over 200,000 Machines, Will Be Weaponized for Years," *Cyberscoop*, April 24, 2017, https://tinyurl.com/y8lgpfmk.

2. Thomas P. Bossert, "It's Official: North Korea Is behind WannaCry," *Wall Street Journal*, December 18, 2017, https://tinyurl.com/y5fas4pz.

3. Garrett M. Graff, "Indicting 12 Russian Hackers Could Be Mueller's Biggest Move Yet," *Wired*, July 13, 2018, https://tinyurl.com/y9s7o8lz.

4. Another latency is also in play insofar as reports claim that the NSA had had the tool for years and alerted Microsoft only when it was stolen. Lily Hay Newman, "The Leaked NSA Spy Tool That Hacked the World," *Wired*, July 3, 2018, https://tinyurl.com/yd774jt8.

5. Bruno Latour, "The Other State of Emergency," *Reporterre*, November 23, 2015, 1.

6. Latour, "The Other State," 1, our emphasis.

7. McKenzie Wark, *The Beach beneath the Street: The Everyday Life and Glorious Times of the Situationist International* (London: Verso, 2011).

8. See, e.g., Richard Grusin, ed., *Anthropocene Feminism* (Minneapolis: Minnesota University Press, 2017).

9. Ian Hodder, "The Entanglements of Humans and Things: A Long-Term View," *New Literary History* 45, no. 1 (2014): 25.

10. See, e.g., Simone Browne, *Dark Matters: On the Surveillance of Blackness* (Durham, NC: Duke University Press, 2015); Nanna Bonde Thylstrup, *The Politics of Mass Digitization* (Cambridge, MA: MIT Press, 2019).

11. Andrew Couts, "Guarding Ground Zero: A Look at the Amazing and Terrifying Artificial Intelligence That Will Protect the New World Trade Center," *Digital Trends*, February 14, 2012, https://tinyurl.com/yy599uh2.

12. PWP Landscape Architecture, "National 9/11 Memorial," accessed May 13, 2019, https://tinyurl.com/y22mdze2.

13. PWP Landscape Architecture, "National 9/11 Memorial."

14. There is an uncanny parallel between the falling water and the disturbing images of the people jumping from the Twin Towers. As documented by Tom Junod, the images of people jumping—including the iconic image of the falling man by photographer Richard Drew—became contested material because they seemed to contradict the heroism of those who died in the collapse, presenting them as if embracing death with a thing-like stoicism that did not match the event's mythologies of lifesaving human action. Tom Junod, "The Falling Man: An Unforgettable Story," *Esquire*, September 9, 2016, https://tinyurl.com/gtfvo7h.

15. John Matson, "Commemorative Calculus: How an Algorithm Helped Arrange the Names on the 9/11 Memorial," *Scientific American*, September 7, 2011.

16. danah boyd, "Social Network Sites as Networked Publics: Affordances, Dynamics, and Implications," in *Networked Self: Identity, Community, and Culture on Social Network Sites*, ed. Zizi Papacharissi (New York: Routledge, 2010), 39.

17. Tarleton Gillespie, "The Relevance of Algorithms," in *Inside Technology: Media Technologies: Essays on Communication, Materiality, and Society*, ed. Tarleton Gillespie, Pablo J. Boczkowski, and Kirsten A. Foot (Cambridge, MA: MIT Press, 2014), 167–194.

18. Blake Hallinan and Ted Striphas, "Recommended for You: The Netflix Prize and the Production of Algorithmic Culture," *New Media and Society* 18, no. 1 (2016): 117–137.

19. Wendy Hui Kyong Chun, "Big Data as Drama," *ELH* 83, no. 2 (2016): 363.

20. Eli Pariser, *The Filter Bubble* (London: Viking, 2011); Shoshana Zuboff, *The Age of Surveillance Capitalism* (New York: Public Affairs, 2019).

21. Mark Weiser, "The Computer for the 21st Century," *Scientific American* 265, no. 3 (1991): 94–104.

22. Mark Weiser and John Seely Brown, "The Coming Age of Calm Technology" (1996), https://tinyurl.com/y65nhsag.

23. See, e.g., Orit Halpern, Robert Mitchell, and Bernard Dionysius Geoghegan, "The Smartness Mandate: Notes toward a Critique," *Grey Room*, no. 68 (2017): 106–129; Ulrik Ekman, ed., *Throughout: Art and Culture Emerging with Ubiquitous Computing* (Cambridge, MA: MIT Press, 2013).

24. Kristin Veel, "Make Data Sing: The Automation of Storytelling," *Big Data and Society* (2018).

25. See, e.g., Clemens Apprich, Wendy Hui Kyong Chun, Florian Cramer, and Hito Steyerl, *Pattern Discrimination* (Minneapolis: University of Minnesota Press, 2019).

26. Bernhard Siegert, *Cultural Techniques: Grids, Filters, Doors and Other Articulations of the Real* (New York: Fordham University Press, 2015), 117.

27. David W. Dunlap, "Looking to a Wall That Limited the Devastation at the World Trade Center," *New York Times*, September 11, 2013.

28. Cavendish Phillips, "Giant Bathtub Will Hold 110-Story Towers," *New York Times*, December 30, 1966.

29. See, e.g., Roberto J. González, "Hacking the Citizenry? Personality Profiling, 'Big Data' and the Election of Donald Trump," *Anthropology Today* 33, no. 3 (2017): 1–20.

30. David Lyon, *Surveillance Society: Monitoring Everyday Life* (Buckingham: Open University Press, 2001), 37–48.

31. Gigasec, "WannaCry 2.0: EternalRock 'Virus' on the Loose! Act!!," accessed May 13, 2019, https://tinyurl.com/y5elfbbs.

Chapter 5

1. Ara Wilson, "The Infrastructure of Intimacy," *Signs* 41, no. 2 (2016): 270.

2. When the mayor of New York City promised to introduce legislation "to ban the glass and steel skyscrapers that have contributed so much to global warming," the *New York Times* was quick to point out that this was unlikely to be more than rhetoric. Jeffrey C. Mays, "De Blasio's 'Ban' on Glass and Steel Skyscrapers Isn't a Ban at All," *New York Times*, April 25, 2019.

3. Hans Ulrich Gumbrecht, *After 1945: Latency as Origin of the Present* (Stanford, CA: Stanford University Press, 2013); Hans Ulrich Gumbrecht, *Broad Present: Time and Contemporary Culture* (New York: Columbia University Press, 2014).

4. Gumbrecht, *After 1945*, 22–23.

5. English translation in Gumbrecht, *After 1945*, 139; original in Martin Heidegger and Shinichi Hisamatsu, "Die Kunst und das Denken: Protokoll eines Colloquiums am 18. Mai 1958," in *Japan und Heidegger*, ed. Hartmut Buchner (Sigmaringen: Thorbecke, 1989), 215.

6. Zoë Sofoulis, "Container Technologies," *Hypatia* 15, no. 2 (2000): 192.

7. Charlie Q. L. Xue, *Building a Revolution: Chinese Architecture since 1980* (Aberdeen: Hong Kong University Press, 2006).

8. Rem Koolhaas, "Beijing Manifesto," *Wired Magazine*, August 1, 2004, https://tinyurl.com/y66gy7kh.

9. Douglas Spencer, "The Architecture of Managerialism: OMA, CCTV and the Post-Political," in *Architecture against the Post-Political: Essays in Reclaiming the Critical Project* (Abingdon: Routledge, 2014), 151–166.

10. Koolhaas, "Beijing Manifesto."

11. Sudjic wonders whether Koolhaas would have been less bitter if "the board of the Whitney had not all chosen to dispense with his services in the course of just twelve months." Deyan Sudjic, *The Edifice Complex: The Architecture of Power* (London: Penguin, 2011), 148.

12. David Harvey, *The Ways of the World* (London: Profile Books, 2016).

13. Rem Koolhaas, "Generic City," in *S, M, L, XL* (New York: Monacelli Press, 1995), 1239–1264.

14. Paul Goldberger, "Forbidden Cities: Beijing's Great New Architecture Is a Mixed Blessing for the City," *New Yorker*, June 30, 2008.

15. Nicolai Ouroussoff, "Koolhaas, Delirious in Beijing," *New York Times*, July 11, 2011.

16. Line Henriksen, Morten Hillgaard Bülow, and Erika Kvistad, "Monstrous Encounters: Feminist Theory and the Monstrous," *Women, Gender and Research*, nos. 2–3 (2017): 4.

17. Donna Haraway, "A Cyborg Manifesto: Science, Technology, and Socialist-Feminism in the Late Twentieth Century," in *Simians, Cyborgs and Women: The Reinvention of Nature* (London: Routledge, 1991), 64–65.

18. William Drenttel, "Koolhaas and His Omnipotent Masters," *Design Observer*, April 9, 2007, https://tinyurl.com/y4zjasz8.

19. Rem Koolhaas, "Junkspace," *October* 100 (2002): 175–190.

20. Rem Koolhaas, "Bigness and the Problem of Large," in *S, M, L, XL* (New York: Monacelli Press, 1995), 496–516.

21. Rem Koolhaas, "Kill the Skyscraper: Beijing CBD Core," in *Content: Triumph of Realization* (Cologne: Taschen, 2004), 471.

22. Karissa Rosenfield, "CCTV Headquarters Named 'Best Tall Building Worldwide,'" ArchDaily, November 8, 2013, https://tinyurl.com/y4mja5mn.

23. David Lyon, *The Culture of Surveillance* (Cambridge: Polity Press, 2018).

24. Wendy Hui Kyong Chun, "Crisis, Crisis, Crisis, or Sovereignty and Networks," *Theory, Culture and Society* 28, no. 6 (2011): 92.

25. Michael Tavel Clarke, *These Days of Large Things: The Culture of Size in America 1865–1930* (Ann Arbor: University of Michigan Press, 2009), 141.

26. Brad Stone, *The Everything Store: Jeff Bezos and the Age of Amazon* (London: Transworld, 2013).

27. Emil Protalinksi, "Amazon Reports $72.4 Billion in Q4 2018 Revenue: AWS Up 45%, Subscriptions Up 25%, and 'Other' Up 95%," *VentureBeat*, January 31, 2019, https://tinyurl.com/y8x83qum.

28. Walter Isaacson, *Steve Jobs: The Exclusive Biography* (New York: Simon & Schuster, 2011), 63. Atari was a prominent games company at the time.

29. "Jack Ma Talkasia Transcript," CNN, April 25, 2006, https://tinyurl.com/y6njd3mj.

30. Jonah Engel Bromwich and Sapna Maheshwari, "Meet the Creator of the Egg That Broke Instagram," *New York Times*, February 3, 2019.

31. See, e.g., danah boyd and Kate Crawford, "Critical Questions for Big Data," *Information, Communication and Society* 15, no. 5 (2012): 662–679.

32. PR Newswire, "Alimama to Launch the First Big Data Platform Based on Real Audiences in the Era of User-Centered Marketing," August 24, 2015, https://tinyurl.com/y58w93vh.

33. Will Wu, "Alibaba Is an Advertising Company, More Than an Ecommerce One," *TLD*, March 22, 2018, https://tinyurl.com/yywy5m2a.

34. Alvin Toffler, *The Third Wave* (New York: Bantam Books, 1980).

35. "I Don't Have Immersion, How Do I Get It?!?!," Duolingo, June 12, 2016, https://tinyurl.com/y4zrppas.

36. Louise Amoore, "Cloud Geographies: Computing, Data, Sovereignty," *Progress in Human Geography* 42, no. 1 (2016): 17.

37. Ben Fox Rubin, "Why Amazon's Dream of a New York City Headquarters Imploded," *CNET*, February 15, 2019, https://tinyurl.com/y33hakmm.

38. See, e.g., Kate Crawford, "The Anxieties of Big Data," *New Inquiry*, May 30, 2014; Henriette Steiner and Kristin Veel, "Living behind Glass Facades: Surveillance Culture and New Architecture," *Surveillance and Society* 9, no. 1 (2011).

39. Susanna Rustin, "If Women Built Cities, What Would Our Urban Landscape Look Like?," *The Guardian*, December 5, 2014.

40. BIG Team, "The Big U," Rebuild by Design, accessed May 13, 2019, https://tinyurl.com/y3ck6bwl.

41. BIG Team, "The Big U."

42. New York City Environmental Justice Alliance, *NYC Climate Justice Agenda: Strengthening the Mayor's OneNYC Plan*, April 2016, 17, https://tinyurl.com/y2jpnvjz.

43. See the film about the project on BIG's homepage, accessed May 18, 2019, https://big.dk/#projects-2wtc.

44. Joanna Zylinska, *The End of Man: A Feminist Counterapocalypse* (Minneapolis: University of Minnesota Press, 2018), 31–32.

45. Frank Donnelly, "At Fresh Kills Landfill, a Heartbreaking Effort after World Trade Center Attacks," *SILive*, September 11, 2011, https://tinyurl.com/yxl5bal5.

Index

Note: Page references in *italics* denote illustrations.

Actor Network Theory (ANT), 139, 151
Adorno, Theodor, 10
Aer nullius, 139–140
Affect theory, 19–20, 60
Ahmed, Sara, 17, 18, 19–20, 151–152
Algorithms, 14, 61, 124, 135, 159–162, 191
Alibaba, 181, 189–190, 191, 192
Alimama, 191
Amazon, 162, 174, 181, 188–189, 192, 194
Amazon Mechanical Turk, 192
"American dream," 71, 104
American independence, 21
Amoore, Louise, 193
Antenna ecology, 13, 177, 183, 192, 201
Anthropocene, 25–26, 122, 139, 140, 151, 195–196, 198
Apel, 135–136
Apple, 135, 189
Arad, Michael, 158, 160
Architectural and digital culture, 4–5, 6–8, 14–16, 22–23, 33
 latent gigantism, as (*see* Latent gigantism)
Architectural history, 6, 14, 86–94, 98
Architectural modernism, 85, 92–93
Architectural montage. *See* Battery Park City; Eiffel Tower

Architectural philosophy, 23
Architectural postmodernism, 24, 73–74, 85–86, 88–89, 111, 112–113, 114, 116, 137, 145, 182. *See also* Twin Towers/WTC
"Artistic" architecture, 37
Art nouveau, 37
Atta, Mohamed, 95–96
Atwood, Margaret, *Oryx and Crake,* 10, 11

Barthes, Roland, 16, 29, 32, 34–35, 47–52, 136
Bartholdi, Frédéric Auguste, 21
Battery Park City (NYC), 79–80, 97–98, 114, 167
Baudrillard, Jean, 16, 73, 96–98, 136, 149–150, 151
Bauhaus, 37, 135
Beijing (China), 7, 181. *See also* CCTV building
"Beijing Manifesto" (Koolhaas), 182, 183–184
Belgian flag, 53, 55
Belgium, 56
Benjamin, Walter, 16, 34–46, 48, 49, 51, 136
 Arcades Project, 36–38, 42–44, 45
Bentham, Jeremy, 119

Berlant, Lauren, 11, 17, 60
Bezos, Jeff, 188
Biases, 14, 18, 22, 24, 163, 179, 191, 195, 202
BIG (Bjarke Ingels Group), 181, 196
Big data, 12, 163, 191
BIG U (Manhattan), *196,* 196–199, 201
Bjarke Ingels Group. *See* BIG
#BlackLivesMatter, 152
boyd, danah, 159
Braun, 135
Broad present, 25, 26, 60, 107, 123, 124, 125–127, 137, 139, 171, 175, 177–178, 202
Brown, John Seely, 162–163
Burj Khalifa (Dubai), 198
Bush, George W., 65

Calatrava, Santiago, 101, 103, 118, 185
Calm technology, 162, 163
Cambridge, Massachusetts (USA), 173, 174
Cambridge Analytica, 195
Carl, Peter, 23
Carpo, Mario, 17, 123–124, 136
CCTV building (Beijing), 7, 21, 180–187, *183,* 195
  competition, 181–182
  latent gigantism, 182–183
Certeau, Michel de, 16, 73, 81–86, 88, 92, 96, 99, 128, 137, 148
Chafer, 147
Charlottesville, Virginia (USA), 144
Chernobyl, 19
Chicago, Illinois (USA), 77, 136
Childs, David, 112, 131–132
China, 6, 181, 182, 183–185, 186, 189, 191, 202
China Central Television. *See* CCTV building
China Media Group (CMG), 181
Chrysler Building (NYC), 132

Chun, Wendy Hui Kyong, 17, 57–58, 59, 98–99, 162, 176, 188
Civicism, 150, 152
Clarke, Michael Tavel, 12, 188
Climate crisis, 11, 60, 139, 149–150, 151, 173, 175–176, 177, 199
  and New York management projects, 181, 196–199
Cloud services, 189, 192–193
CN Tower (Toronto), 112
Cold War, 74, 137
Colonialism, 20, 22, 83, 122
Commonality, 7, 22, 57, 148, 149, 169–170, 179, 180, 201
"Commonness," 140
Condé Nast, 118
Consumer society, 10, 60, 107, 182
Containers, 17, 18–19, 117, 170, 178. *See also* Leaks
Containment, 1, 19, 38, 75, 165–166, 178, 179, 180, 191, 197, 198, 201
Cooper Union for the Advancement of Science and Art (NYC), 134
COP21, 149
Copenhagen (Denmark), 1, 53, 65, 101–102, 103, 126, 173–174, 203
Corner, James, 196, 200
Coulthard, Glen, 140
Council for Tall Buildings and Urban Habitat, 187
Crip theory, 152
Crisis, 1, 22, 46, 56, 59, 60, 68, 70, 71, 140, 180, 188, 201. *See also* Climate crisis; Financial crisis
  of "greatness," 20
  perpetual state of, 58–61, 98–99, 185
Critical regionalism, 89
Cruel optimism, 11, 60, 126, 177, 188
Crutzen, Paul, 26
Cryptocurrency miners, 146–147
Cubist painting, 43
Cultural theory, 4, 6, 7, 11, 15, 16–17, 19, 22, 24–27, 35, 49, 51–52, 72–73,

88, 136–137, 138, 149, 152, 177–178, 181
Cyberespionage, 147
Cyborgs, 52, 84, 186

Data centers, 164, 192
Datafication, 4–5, 116, 119, 150
Data mining, 57, 58, 179, 191
Data profiling, 57
Data storage, 18, 25
Decolonial theory, 150, 151
Denmark, 173
D'Espezel, Pierre, 42
Dichotomies, 4, 13, 16–17, 18, 19, 25–26, 51–52, 81, 82, 146, 179
Dickens, Charles, *A Tale of Two Cities*, 4, 14
Digital and architectural culture. *See* Architectural and digital culture
Digital architecture, 134–135, 136
Digital publics, 12, 119, 124, 152–164, 169–170
Digital technology, 25, 111, 123, 136, 179. *See also* Cloud services; Tech giants
Dot-com bubble, 138
Drenttel, William, 186
Dualisms, 15–16, 18, 51, 52, 187
Dubai, 198
Dubech, Lucien, 42

East Side Coastal Resilience Project, 198
Economic recession, 59
Edison, Thomas, 39
Eiffel, Gustave, 21, 38–40, 48
Eiffel Tower (Paris), 6, 20, 21, 22, 29–62, *30, 34, 39, 44, 47, 54,* 107, 115, 124, 135, 195, 200
  architectural montage, 31, 33, 35, 36, 43–47, 48–49, 61–62, 164
  digital reincarnations, 56–57, 61
  Peace for Paris, 53, *55,* 56, 61
  gigantism, 23, 33–35, 46–47

latent, 24, 33, 52–53, 57, 59, 61–62, 73
  semantic, 33, 45, 48, 49, 51, 53, 61
  vertical, 33, 42, 48, 56
illumination, 32, 53, 55–56, *60*
"incomparableness," 42
iron construction, 37, 38–39, 40, 42, 87
originating idea, 38–39
radio transmission, 5–6, 14, 35, 43, 164
restaurant Le Jules Verne, 29, 31
as skyscraper, 164
symbolism, 42, 49, 158, 176
terror attacks, and, 32–33, 53–56, 176, 179
World's Fair 1889, *40, 41, 50,* 80
Emerson, Ralph Waldo, 88
Empire State Building (NYC), 104, 132
Engelke, Peter, 26
"Engineering" architecture, 37
Entrapment, 3, 17, 27, 108, 139, 151–152, 180, 199, 201
EternalBlue, 146–147, 170, 171
EternalRocks, 170–171
Europe, 32, 37, 49, 56, 62, 74, 115
European Union, 2–3
Expandability, 9

Facebook, 12, 29, 31, 56–57, 176, 179
Fancy Bear, 147
FarmVille, 176
Feminist science fiction, 186
Feminist theory, 10, 12, 17, 18, 52, 71–72, 74, 83–84, 140, 150, 151
Field Operations, 200
Financial crisis (2008), 59, 184–185
Finn, Maria, drawings by, x, xiv, 5, 8, 21, *28,* 33, *64,* 71, *142,* 158, *172*
Flammarion, Camille, *Thunder and Lightning,* 34
Flattening, 9, 10, 11, 12, 19–20, 22, 23, 25, 26, 53, 57, 60, 61, 108, 127–128,

Flattening (cont.)
131, 135, 137, 139, 140, 147, 151, 159, 161, 163, 171, 179–180, 186, 189, 201. *See also* Ontological slippage
Foucault, Michel, 119
France, 2, 32, 149, 150. *See also* Eiffel Tower; Paris
Frankfurt School, 49
French poststructuralism, 49
French Revolution, 21, 56
Fresh Kills Park, Staten Island (USA), 7, 199, 201, *201*
 landfill, 196, 199, *200*
Friedell, Egon, 38
Friedland, Sarah, 57–58
Fukuyama, Francis, 137

Gardner, Anthony, 111
Gender, 17–19, 22, 51–52, 57, 71, 72, 84, 92, 99, 140, 163, 191, 195, 202
 female imaginary, 186
 phallic imaginary (*see* Phallic structures)
Germany, 106, 178
Gigantism, 3–5, 8–13, 20–21, 22, 23
 definition, 4, 8, 112
 digital infrastructures, 12
 forms of, 4–5, 13–17 (*see also* Horizontal and vertical gigantism; Latent gigantism; Linear gigantism; Semantic gigantism)
 historical context, 23
 nostalgic notions of, 22
 phallic imaginary (*see* Phallic structures)
 sameness, 10
 tower buildings, 12, 13, 14, 15, 17–18 (*see also* Eiffel Tower; One World Trade Center; Twin Towers/WTC)
Gigasec Services, 170–171
Gillespie, Tarleton, 161–162
Givan, Rebecca Kolins, 194

Global communications infrastructures, 5, 71, 77, 95
Global energy consumption, 192
Global geography, 164
Globalization, 187
Global networks, 75, 152
Global recognizability, 189
Global resonance, 22
Global roaming, 123–124
Global routing, 70
Global village, 185, 191
Global warming, 149, 151, 173, 175. *See also* Climate crisis
Google, 70, 99, 194
Gothic architecture. *See* Neo-Gothic architecture
Great acceleration, 26
Greatness, crisis of, 20
Greek mythology, 8, 10, 186, 188
Gropius, Walter, 136
Ground Zero site (NYC), 6, 19, 99, 111, 148, 150, 179
 flooding, *197*
 Memorial Museum (9/11), 153, *154*, 166, 168
 memorial site, 108, 112, 118, 122, 148, 152–164, *153*, *154*, *155*
  garden of trees, 155–158, *157*
  pools, 152, 158–159, 162
  victims' names on bronze panels, 159–162
 naming of, 106
 rebuilding (*see also* One World Trade Center)
  design competition, 22, 181, 182
  proposed Freedom Tower, 129, 131
  replacement tower, 106, 109, 111–112
 securitization, 153, 188
 slurry wall, 166–168, *167*, *169*, 171, 179
 train station, 168–169
Gumbrecht, Hans Ulrich, 25, 177–178, 185, 186

Index

Hacking tools, 19, 146–147, 170
Haraway, Donna, 10–11, 13, 16, 26, 74, 83–84, 118–119, 148, 200
  "A Cyborg Manifesto," 52, 84, 99, 186
Hartman, Saidiya, 17, 138, 139, 141
Harvey, David, 183–185
Hayles, N. Katherine, 26
Heidegger, Martin, 4, 13, 16, 18, 23, 112, 178
Highmore, Ben, 84
History
  architecture and, 86–94, 112, 147
  concepts of, 18, 23, 24, 35, 36, 45, 46, 59, 107, 114, 116, 122–124, 137
  "end of history," 137, 138
  "historical time," 24
Hodder, Ian, 17, 151–152
Hollande, François, 149
Hong Kong, 2
Horizontal and vertical gigantism, 13–17, 24, 84, 96, 114, 188, 192, 198
  Eiffel Tower, 33, 35, 42, 45–46, 47, 48, 49, 51, 56, 58, 61
  One World Observatory, 105–107, 128, 136–137, 147
  One World Trade Center, 136–137
  World Trade Center (*see* Twin Towers/WTC [NYC])
Horkheimer, Max, 10
Hudson River (New York), 19, 79, 166, 168, 171
Hunters and gatherers, 18
Hurricane Sandy, 168, 171, 196, *197*
Huxtable, Ada Louise, 93, 143, 146
Hyperobject, notion of, 11–12, 26, 189

Inequalities, 4, 8, 9, 12, 56, 72, 83, 92, 95, 140, 141, 152, 163, 179, 180, 187, 191
Information superhighway, 14
Ingels, Bjarke, 197, 198
Instagram, 12, 127, 191

International Chamber of Commerce, 74
Internet, 14, 19, 57, 59, 68–71, 162–163, 192
Internet Machine, *193*
Internet under Crisis Conditions: Learning from September 11 (2003), 68–71
IPad/iPod, 135
IPhones, 135, 136, 138–139
Iran, 147
Iron construction, 37, 38, 42, 43, 135, 136. *See also* Eiffel Tower
Irony, age of, 88–89, 91–92, 106, 107, 114, 145
ISIS (Islamic State of Iraq and Syria), 56, 61
Isozaki, Arata, 182
Italy, 123

Jencks, Charles, 111, 112, 182
Jenner, Kylie, 191
Jewish Museum Berlin, 109, 111
Jewish mysticism, 36, 46
Jobs, Steve, 189
Jugendstil, 37, 38
Jullien, Jean, *Peace for Paris,* 53, 55, 56, 61

Kalevala, 203
Koechlin, Maurice, 38–39
Koolhaas, Rem, 22, 180–188, 195, 197
  "Beijing Manifesto," 182, 183–184
Koselleck, Reinhart, 24, 59

Latent gigantism, 22–27, 52, 72, 73, 123, 124, 136, 147, 152, 163, 171, 173, 175–181. *See also* Flattening; Ontological slippage
  ambient form, 26
  architectural and digital culture, 5, 22, 24–25, 148, 152, 176–177, 180, 191
  China, 182–183, 185, 186, 187
  climate crisis, and, 176, 196, 199

Latent gigantism (cont.)
  commonality and publicness, 139, 140, 141, 146, 148, 169–170
  meaning of, 11, 24–25
  paradoxes, 7, 11, 24, 25, 117, 134, 177, 180, 199, 201–202
  tech giants, 188–189, 191–192
  temporality, 137, 138–139, 171, 175–176
  towers (see Eiffel Tower; One World Trade Center)
  Trump's Twitter comments, 169
  world_record_egg, 191
Latour, Bruno, 16, 139, 148, 149–51, 152
Leaks, 19–20, 26–27, 52–53, 136, 151, 152, 162, 163, 164–171, 173, 175, 178, 179, 180, 183, 201
  crisis, and, 59–61
  Eiffel Tower, 61–62, 164–165
  EternalBlue ("fifth leak"), 146–152
  new media, 57–58, 61
  One World Trade Center, 120, 124, 129, 162, 165–166
  slurry wall in 9/11 Memorial Museum, 166–168
  WikiLeaks, 195
Le Guin, Ursula, 18
Libeskind, Daniel, 109–12, *110*, 116, 122, 129, 131, 132, 137, 168
Lin, Maya, 160
Linear gigantism, 24, 25, 33, 45–46, 61, 72, 73, 105, 132, 185, 187
  New York City, 75
  tech giants, 188–189, 192
  Twin Towers, 92–93, 99
Linear history, 124, 126
Linearity, 14, 22, 23, 24, 35, 59, 60, 73, 123, 138, 168, 188–189
"Livable city," 115
Lyon, David, 170, 188
Lyotard, Jean François, *The Postmodern Condition*, 137

MacBook, 135
Machine learning algorithms, 14, 191, 202
"Make America Great Again" (slogan), 20
Malm, Andreas, 26
Malmö (Sweden), 6, 101, 103, 118
Manhattan (NYC), 7, 66, 73, 79, 81–82, *82*, 83, 120–122, 125, 127, 128, 199
  BIG U project, *196*, 196–199
  skyline, 81, *105*, 198
  skyscrapers, 5, 84–85, 104, 117–118 (see also One World Trade Center; Twin Towers/WTC)
Marne, battle of (1914), 35
Marxism, 36, 46, 137, 150
Matrix (film), 95
Mattern, Shannon, 17, 23
Maupassant, Guy de, 47
Ma Yun (Jack Ma), 189, 190, 191
McNeill, J. R., 26
Media theory, 159, 161, 164
Megadams, 9
Metadata, 120
#Metoo, 152
Metropolitan mainstream. See New metropolitan mainstream
Meyer, Alfred Gotthold, 44
Microsoft, 146
Mies van der Rohe, Ludwig, 135–136, 164
Modernism, 85, 89, 92–93, 137. See also Postmodernism
Morton, Timothy, 11–12, 26, 185
Mumford, Lewis, 18

National Research Council, 68
National Security Agency (NSA), 146, 171
Nazism, 106
Neo-Gothic architecture, 73, 85, 86, 88, 93, 124, 156
Netflix, 162

Index

Networked publics. *See* Digital publics
New metropolitan mainstream, 107, 112, 114–116, 135, 136–137
New York City, 7, 21, 66, 75–76, 79–80, 82, 115, 136, 194. *See also* Manhattan; One World Trade Center; Twin Towers/WTC
  climate-change management, 181, 196–199
  Hurricane Sandy, 196, *197*
  superconnected node, 69
  terror attacks (9/11/2001) (*see* September 11 attacks)
New York City Environmental Justice Alliance, 197–198
New York City Panel on Climate Change, 196
New York Times, 71, 93, 143, 166, 191
Ngai, Sianne, 16
Nice (France), terror attacks (2016), 29, 31
Nigeria, 170
Nixon, Rob, 9, 17
North Korea, 146

Observatories. *See* Eiffel Tower; One World Observatory
Occupy Wall Street movement, 150
Olympic Stadium (Beijing), 181
OMA (Office for Metropolitan Architecture), 181, 182, 197
One World Observatory, 6, 101–108, 118–129, 131, 136, 139, 197, 200
  horizontal and vertical gigantism, 105–107, 128, 136–137, 147
  mediated experience
    elevator ride, 118–119, 120–123, *121*
    observatory space, 124–129, *127, 128, 130,* 171, 179
One World Trade Center (NYC), 3, 5, 7, 20, 103, 104–108, *105, 133,* 147, 152, 153, 165–166, *165,* 179, 188, 195, 198, 200

  architecture, 5, 6, 106, 107–108, 112, 119, 122, 131–132, 134–136, 179, 182
  competition, 109, 181
  initial design proposal (2003), 109–112, 116, 122, 129, 131, 132, 137
  postmodernism, 112–113
  container qualities, 105, 117, 147, 166
  horizontal and vertical gigantism, 136–137
  latent gigantism, 5, 23–25, 73, 103, 105, 106, 108, 116, 122, 124, 131, 132, 134, 136, 137, 147–148, 179
  mechanical floors, 117
  memorial park and pools (*see* Ground Zero site)
  new metropolitan mainstream, and, 109–116, 135, 136–137
  observatory (*see* One World Observatory)
  securitization, 116–118, 119, 120, 132, 143, 147, 166
  semantic gigantism, 132
  size of, 5, 104
  Statue of Liberty, and, *131, 132*
  subway station and shopping facility, 118, 125
  tenants, 117–118
  tourist site, 107, 108, 134
  transparency, 166
Ontological slippage, 16, 17, 19, 23, 25, 26, 61, 108, 137, 171, 179, 185

Panopticon, 85, 119
Parallelism, 17, 108, 151
Paris (France), 21, 36, 46, 47–49, 115. *See also* Eiffel Tower
  Place du Trocadéro, *47*
  terror attacks (2015), 21–22, 32–33, 35, 53, 148, 149, 176, 179
Pentagon, 96
Petit, Emmanuel, 88–92, 93, 106–107
Pettman, Jan Jindy, 72

Phallic structures, 17, 18, 19, 36, 51, 72–73, 83, 86, 87, 103, 105, 118, 132, 147, 186
Philosophical postmodernism. *See* Postmodernism
Posthistory, 93, 137
Postmodernism, 107, 123, 137
　architectural (*see* Architectural postmodernism)
　epistemologies, 10, 74, 84, 124
　philosophical, 73–74, 85–86, 92, 137
Poststructuralism, 24, 49, 89, 137
Pratt, Geraldine, 72
Predictive analytics, 57, 179
Pritzker Architecture Prize, 181–182
Pruitt-Igoe buildings, St. Louis
　demolition of, 89, *90,* 98

Queer theory, 17, 52, 150

Race, 10, 22, 26, 57, 138, 140, 152, 163, 191
Rams, Dieter, 135–136
Ransomware, 146–147, 171
Rasmussen, Mikkel Bolt, 138–139
Reid, Julian, 94
Relational thinking, 10, 52, 84, 139, 148, 149, 150, 151 (*see also* Utopianism)
Resilience, 42, 166, 196, 198–199
Rosner, Victoria, 72
Runia, Eelco, 178, 185
Russia, 147

St. Louis, Missouri. *See* Pruitt-Igoe buildings
Scaffolding, 1–3, 4, 7, 8, 43, 45, 120
Scalability, 9
Schmid, Christian, 107, 112–113, 115
Scott, Fiona, 195
Sears Tower (Chicago), 77
Seattle (USA), 194

Securitization. *See* Ground Zero site; One World Trade Center; Surveillance
Semantic gigantism, 23, 24, 25, 33, 45, 48, 85–86, 91–92, 99, 105, 185, 187, 192. *See also* Eiffel Tower; Twin Towers/WTC
September 11 attacks (9/11/2001), 5, 14, 21–22, 32, 56, 65–74, *69,* 80, 94–99, *96,* 106, 116, 149–150, 159, 176
　historicist interpretation, 89, 91
　Memorial Park and Museum (*see* Ground Zero site)
　scales of impact, 68–74
　wakeup call, as, 137–138
Seven World Trade Center (NYC), 79, 96, *165*
Shadow Brokers, 146, 171
Siegert, Bernhard, 164, 170
Skidmore, Owings, and Merrill, 112, 131, 135, 182
Skyscraper of the Year Award, 187
Skyscrapers, 9, 14, 81–83, 84, 114–115, 135, 164
　phallic imaginary (*see* Phallic structures)
　"wombs with a view," 18, 73, 86, 92, 94, 123, 132, 147, 166
Slurry wall. *See* Ground Zero site
Smart city, 98, 198
Smartphone, 136
Smart technologies, 18, 163
Snodgrass, Eric, 56–57
Snowden, Edward, 195
Social media, 6, 22, 32, 53, 55, 56, 57–59, 98, 99, 116, 119, 120, 127, 148, 159, 166, 169, 179, 194
Sofoulis, Zoë, 17, 18–19, 73, 86–87, 178
Staten Island. *See* Fresh Kills Park
Statue of Liberty, 21, 71, 109, 129, *131,* 132, *132*
Stewart, Susan, 16

Index   231

Stickiness, 19, 192, 199, 200
Sticky entrapment, 151–152
Stockhausen, Karlheinz, 97, 99, 106
Studio Libeskind, 109
Sudjic, Deyan, 182
Surveillance, 57, 58, 81, 119, 147, 153, 170, 182–183, 187, 188, 194
Sweden, 101

Tech giants, 6, 181, 188–195. *See also* Amazon; Apple
Temporality, 59, 70, 93, 104, 161, 168, 198
  latent gigantism, 138, 139, 171, 175–176
  One World Observatory, 120, 122, 123
  of resilience, 199
  of revolution, 138
Terra nullius, 140
Terror attacks, 21–22, 29, 31, 32–33, 35, 149–151. *See also* Paris; September 11 attacks
Titanic, 19
Todd, Zoe, 139–140
Tribeca (NYC), 79, 198
Trump, Donald J., 20, 143, 144, 146, 169
Trump, Ivanka, 144
Trump Tower (NYC), 143–146, *145*
Tsing, Anna Lowenhaupt, 4, 9, 13, 17
Turkey, 29
Turning Torso (Malmö), 101, 118, 185
Twin Towers/WTC (NYC), 3, 6, 20, *76, 77, 78, 92,* 104, 116, 150. *See also* World Trade Center
  architecture of, 6, 21, 74, 85, 86–94, 98, 99, 124, 182
  facades, 164–165
  postmodernism, 93–94, 106, 109, 111, 137
  construction, under, *91*
  containers, as, 86–87

  destruction in terror attacks (9/11/2001), 5, 6, 32, 56, 65–74, *69,* 92, 94–99, *96,* 106, 110, 137–138 (*see also* September 11 attacks)
  gigantism, 72–76, 77, 79, 80, 93–94
  horizontal and vertical, 66, 69, 72, 74–80, 81–82, 83, 85, 87, 93, 95, 98, 106
  semantic, 23, 72, 86, 111, 131
  media history, 98–99
  North Tower, 112, 132
  radio and transmission tower, 77–78
  phallicism, 72–73, 87
  rebuilding, 97–98 (*see also* Ground Zero site)
  slurry wall (*see* Ground Zero site)
  South Tower, 79, 120
  observation deck, 81, *82*
  utopianism, 74, 81, 82–83, 92, 99
  view from above, 80–86, *82*
Twitter, 53, 59, 169
Two World Trade Center (NYC), *82,* 198

Ubiquitous computing, 162, 163
Ulm School of Design, 135
United Airlines Flight 175, 69
United Nations Climate Change Conference (COP21), 149
United States
  Amazon headquarters, 194
  crisis of "greatness," 20
  foreign policy, 74, 94–95
  presidential election (2016), 147, 162
United States Customs and Border Protection, 118
Urbanization, tendencies of, 115–116
Urban planning, 7, 14, 89, 105, 115, 125, 140. *See also* Battery Park City; New metropolitan mainstream
  New York climate management projects, 181, 196–199
Urbs nullius, 140

Utopianism, 4, 9, 10, 12–13, 22, 52, 149, 150, 152, 187
   World Trade Center, 74–75, 81, 82–83, 92, 99

Verizon building (NYC), 68–69
Vertical gigantism. *See* Horizontal and vertical gigantism
Vesely, Dalibor, 43
Victorian culture, 37–38, 40
Vietnam Veterans Memorial (Washington, DC), 160
Virilio, Paul, 73, 80–83, 88, 96
Vogue, 118

Walker, Peter, 155, 157
WannaCry, 146, 170, 171
War on Terror, 138, 181
Weibel, Peter, 119
Weiser, Mark, 162–163
Western industrial culture, 4, 6, 16, 26, 46, 179, 199
   gigantism and, 9, 12, 20–21, 24, 35, 73, 181, 195
   horizon of expectation, 59
Wigley, Mark, 13, 177, 201
WikiLeaks, 195
Wilson, Ara, 177
Windows operating system, 146, 147
Wired, 118
Wittenberg, David, 12
Wombs, 18, 116–124, 146
  with a view (*see* Skyscrapers)
world_record_egg, *190*, 191
World's Fair (New York, 1939), 74, 86
World's Fair (Paris, 1889), 35, 38, 42, 80, 158
World Trade Center (NYC), 5, 6, 23, 76, 77, 89, *92*, 93, *156*. *See also* Battery Park City; Ground Zero site
  rebuilding (*see* One World Trade Center)
  terror attacks (9/11) (*see* September 11 attacks)
  towers (*see* Twin Towers/WTC)
  vision of, 73, 74–75, 76, 79–80
  World's Fair pavilion (1939), 74, 86
World Trade Center United Family Group, 111
World Trade Organization, 181
World War II, 74, 106, 178
Worldwide web, 14

Xerox PARC, 162
Xue, Charlie Q.L., 181

Yamasaki, Minoru, 75–76, 88, 89, 93, 97, 109, 166–167
YouTube, 59, 124
Yussof, Kathryn, 26

Žižek, Slavoj, 16, 73, 94–95, 106, 114
Zylinska, Joanna, 17, 140, 198